STUDIES IN CONVECTION

Volume 2

STUDIES IN CONVECTION

Volume 2

Theory, Measurement and Applications

Edited by

B. E. LAUNDER

University of California, Davis

1977

ACADEMIC PRESS

London New York San Francisco

A Subsidiary of Harcourt Brace Jovanovich, Publishers

ACADEMIC PRESS INC. (LONDON) LTD.
24/28 Oval Road,
London NW1 7DX

United States Edition published by
ACADEMIC PRESS INC.
111 Fifth Avenue
New York, New York 10003

Library of Congress Catalog Card Number: 75-13621
ISBN: 0-12-438002-6

Printed in Great Britain by
Whitstable Litho Ltd., Whitstable, Kent.

PREFACE

Of the many fields in which convection processes play an important role that of Combustion is probably the one which, in the last few years, has experienced (and continues to experience) the greatest resurgence of interest. The expansion of activity springs directly from two of the major issues of the 1970's: the "Energy Crisis" and what may be called "Concern for the Environment". To extend the lifetime of remaining oil reserves it is widely recognized that more efficient means must be devised for converting hydrocarbon fuels to usable energy. Yet, at the same time, those with proper concern for the quality of the air we breathe decree that the concentration of harmful biproducts of combustion – such as the oxides of nitrogen – should be reduced to negligible proportions. The satisfaction of both desiderata poses the most formidable research challenge over the next ten years.

In recognition of the growth and interest in combustion research the second volume of Studies in Convection is devoted entirely to aspects of convection relating to chemical reaction. While necessity is regarded as the mother of invention, the offspring's progress is undeniably helped by having a father on the Board of Directors. In this respect Combustion, as an area of research, seems to have been fortunate for progress over the preceding decade, largely on non-reacting flows, has thrown up methods of analysis, new instrumentation and a general understanding of turbulence that are being brought effectively to bear on combustion phenomena. This feature emerges again and again throughout the book.

As with Volume 1, the four articles which form the present volume include both experimental and theoretical approaches. Each article is designed to provide a synthesis of several years' research by its author and his colleagues. In compiling them we have tried to provide up-to-date and readable surveys that will prove valuable both to the newcomer and the experienced researcher. If the volume has succeeded all credit must go to Professor Becker, Dr. Elghobashi, Professor Libby and Professor Pratt; for their cooperativeness and patience I am deeply thankful.

Davis, California
June 1977

BEL

CONTENTS

STUDIES IN VARIABLE-DENSITY AND REACTING TURBULENT SHEAR FLOWS

Paul A. Libby

Department of Applied Mechanics and Engineering Sciences
University of California at San Diego

ABSTRACT

Experimental research on variable density flows such as arise in the turbulent mixing of helium and air and theoretical research on turbulent reacting flows are described. Extended hot-wire anemometry is shown to provide interesting and useful data on variable density effects in turbulence. Because reacting flows of practical interest involve significant heat release and therefore significant density variations, there is a close connection between our experimental effort and our study of turbulent reacting flows. The theoretical treatment of the case of non-premixed reactants involving fast chemistry both with and without heat release is discussed. It is concluded that turbulent flows involving variable density and chemical reactions are rich in challenging problems.

1. INTRODUCTION

1.1 Scope of the Article

The writer's research in recent years has involved theoretical and experimental investigations related to two aspects of turbulent shear flows: the effects of variable density and of chemical reactions in turbulent flows. These two aspects are interrelated since practically interesting chemical reactions usually involve sufficient heat release for appreciable density variations to arise. However, each of these topics in turbulence is sufficiently rich in fruitful problems for useful separate studies to be carried out.

The present article reviews the recent work undertaken in this area by the writer and his colleagues. Section 2 is concerned with the measurement of density and velocity fluctuations using hot-wire anemometry while Sec.3 considers some of the problems associated with turbulent combustion in the limit of infinitely fast chemical reaction. First, however, some of the main problems and concepts are briefly reviewed.

1.2 Flows with Variable Density

Variations in density can arise in turbulent flows for a variety of reasons. For many years the turbulent boundary layer under high-speed conditions such as occur in many aerospace applications has been the subject of intensive investigation because of its importance in determining frictional drag, heating rates and equilibrium temperatures in re-entry and high-speed flight. The most complete assessment of our understanding of compressible turbulent boundary layers is probably provided by the proceedings of a NASA conference on this subject [1]. From these it appears that, for supersonic and slightly hypersonic boundary layers with modest pressure gradients and without the injection of foreign species, our understanding of the gross behaviour of these flows is satisfactory. When either these limitations is violated, or when details of turbulence correlations and fluctuations are sought, our confidence is diminished.

This assessment of our knowledge of gross behaviour reflects in large measure the useful concept proposed by Morkovin [2], generally called the "Morkovin hypothesis". He suggested that if the Mach number of the turbulent fluctuations is much less than unity (a situation believed to be true in turbulent boundary layers with edge Mach numbers less than 5 or 6 and in jet flows with rather smaller Mach numbers – perhaps 1.5 to 2), the effects of density fluctuations could be neglected. The direct effect of variable density via variations in mean density must still be taken into account but the Morkovin hypothesis provides a convenient justification for neglecting a wide variety of terms associated with density fluctuations.

For even low speed turbulent flows variations in density can arise due to heterogeneities in temperature and composition. There are well-known, simple manifestations of density effects in low speed flows. For example, consider the case of a circular jet discharging into a quiescent medium. If the jet fluid and the surroundings consist of the same fluid and are at the same temperature, the jet will spread and its velocity will decay at certain rates. If the jet fluid has a lower density, either because of temperature or composition, the spreading rate and the rate of decay of the jet velocity are increased. The opposite effects are observed if the jet fluid has a higher density than the surrounding medium. The gross behaviour of such variable density jets close to the jet exit are explained by the concept of an equivalent orifice diameter based on consideration of total jet momentum (cf. Hinze [3]). Far downstream, the density of all jets becomes that of the surrounding medium as density variations dissipate. Accurate treatment of such a simple manifestation of variable density does not seem to be possible at the present time even with rather sophisticated predictive methods. The reader is referred to Test Case 12 of [4] in this regard. Furthermore, understanding of the effects of density variations on the physics of turbulence, e.g., on entrainment through the interface involving a density difference as well as a vorticity difference, are lacking.

In turbulent flows involving chemical reaction there is in general a close, and poorly understood interaction between the chemical behaviour, and the turbulent mixing processes. However, theoretical considerations suggest that this interaction must depend strongly on variations in density, certainly in mean density, perhaps on density fluctuations as well. Although pressure fluctuations lead to density fluctuations, this effect is usually considered secondary (except, of course, in studies of noise generation) to the influence of temperature and composition on density.

Experimental techniques for variable density flows

It is widely accepted that the most significant advances in our understanding of turbulence arise from a close interplay between theoretical and experimental efforts. It is worthwhile therefore to consider the current status of experiment for turbulent flows with variable density. The hot-wire anemometer has been the most useful tool in experimental turbulence research for fifty years. It and the related sensor, the cold-wire which permits fast-response thermometry, combined with modern techniques of digital analysis have provided the great bulk of data on turbulent flows with essentially constant density.* Some years ago Kovasznay (see [5,6]) showed how the hot-wire could be used to provide data in high speed flows. Recently, Sandborn [9] has reviewed turbulence measurements in compressible flow from hot-wire and laser anemometers and concluded that in certain regions of the boundary layer these devices give reasonably accurate values of the mean shear stress but that near the wall difficulties arise. Optical techniques based on various applications of the laser have in recent years begun to provide a valuable tool for the study of complex turbulent flows, including those involving chemical reactions with many species. Some of these techniques are discussed by Professor Becker in the second article in this volume while Professor J.H. Whitelaw has discussed various applications of the laser-Doppler anemometer in Volume 1 of this series [10]. A number of groups have used these methods to obtain measurements of velocity and species fluctuations individually; however, the techniques for obtaining by optical methods the interesting *cross*-correlations between velocity and concentrations are only just beginning to emerge [10,11].

Without question there is still a great need for a variety of high-quality data relating to many turbulence quantities in order to clarify our ideas on turbulent flows with significant density variations. Since such density variations can arise from many causes and in many types of flows, various techniques are called for. There is always the hope, of course, that information on density effects in one type of flow will shed light on similar effects in other flows. For example, it is to be hoped that information on the low speed turbulent mixing of helium and air, a case in which the density effects are associated with heterogeneities in composition, may be useful in understanding the high speed turbulent boundary layer in which the heterogeneities in temperature due to viscous heating cause the density variations.

The phenomenology of turbulent flows with variable density

When we consider the phenomenology of turbulent flows with variable density, we find interesting differences of strategy in the treatment of the density. In what might be termed the straightforward approach, the density is treated as all other variables; that is, decomposed into a mean and fluctuating component. Thus the inertia terms in the momentum equation give rise to

---

* Gibson [7] and van Atta [8] provide recent reviews of modern developments in experimental turbulence involving digital techniques. See Bradshaw [6] for *inter alia* a survey of hot-wire anemometry.

$$\overline{\rho u_1 u_2} = \bar{\rho}\bar{u}_1\bar{u}_2 + \bar{\rho}\overline{u_1' u_2'} + \bar{u}_1\overline{\rho' u_2'} + \bar{u}_2\overline{\rho' u_2'} + \overline{\rho' u_1' u_2'} \quad (1)$$

where we use customary notation. The first term on the right is the usual one kept on the left side of the equations; the second is the variable-density analogue of the usual Reynolds stress; the next two are direct contributions to the momentum flux via an interaction between density and velocity inhomogeneities. The final term is a triple correlation which can probably be neglected. Morkovin [2] labels the third and fourth terms on the right side "mass transfer" terms; these are the ones neglected when the Morkovin hypothesis is invoked. We see that when density fluctuations must be taken into account, a variety of additional terms need to be evaluated either as direct dependent variables or by modelling so that they may be expressed in terms of other dependent variables.

There are alternative approaches to the treatment of density. The most useful is due to Favre [12]* according to which the density is not decomposed but all flow variables except the pressure are mass averaged. Thus equation (1) becomes without approximation

$$\overline{\rho u_1 u_2} = \bar{\rho}\tilde{u}_1\tilde{u}_2 + \overline{\rho u_1'' u_2''} \quad (2)$$

where, for example, $u_i = \overline{\rho u_i}/\bar{\rho} + u_i'' = \tilde{u}_i + u_i''$. Because of the mass averaging $\overline{u_i''} \neq 0$ but $\overline{\rho u_i''} = 0$. From equation (2) we see that the right side now consists of only two terms; the first is the usual mean term which remains on the left side of the momentum equation while the second is the mass-averaged Reynolds stress term.

Comparison of equation (1) and (2) indicates a dramatic simplification if Favre averaging is used. However, several notes of caution are called for; the Reynolds stress term represents a new unknown which is introduced by the averaging process and which must be either treated as a new dependent variable or modelled in terms of other primary dependent variables. Suppose the latter course is followed, e.g., by invoking an eddy transport notion so that $\overline{\rho u''v''} = -\bar{\rho}\nu_T \partial\tilde{u}/\partial y$. There is a strong tendency simply to carry over to variable density flows the models which have served so well in turbulent shear flows with constant density without critically assessing the validity and accuracy of doing so. In the example cited earlier the variation of ν_T might be casually taken either to be constant or to vary across a shear layer as in constant density flows. This might be unsatisfactory.†

A second concern arising with Favre averaging relates to comparison with experimental results. If such results are in terms of conventionally averaged quantities, e.g., of $\bar{u}_i$, and the predictive theory yields $\tilde{u}_i$, then to make the comparison there must be used the equation

$$\tilde{u}_i = \bar{u}_i + \overline{\rho' u_i'}/\bar{\rho} \quad (3)$$

* The approach has been extended to the transport equations for the Reynolds stresses by Rubesin and Rose [13].

† Professor K.N.C. Bray has pointed out to the author that the body-force terms in the conservation equation for turbulent kinetic energy, a term of significance in stratified turbulent flows, disappears explicitly with Favre averaging. Presumably the effect of body forces on the turbulent kinetic energy must be hidden elsewhere. This is an illuminating example of the possible deception in the simplicity afforded by Favre averaging.

Thus, before the comparison can be made one of the mass transfer terms of equation (1) must be evaluated. It will be shown that that the differences between conventional and Favre averaged quantities are sometimes small and thus that in these cases only quite accurate data warrant correction before comparison with theory.

These brief comments on turbulent flows with variable density have attempted to convey an impression of an aspect of turbulence which is of great practical importance and rich with interesting, largely unsolved problems. The main efforts of the writer and his colleagues in this area have been aimed at extending hot-wire anemometry to permit detailed, high-quality measurements in well-controlled flow situations involving significant density variations. In particular we have studied the mixing of helium in air with the hot-wire measuring simultaneously, with good time and space resolution, one or more velocity components and helium concentration. As we shall see later the key to successful measurement resides in the probe itself.

1.3 The Situation Concerning Turbulent Reacting Flows

Turbulent reacting flows appear in a wide variety of practical applications. The effect of turbulent mixing on the chemical behaviour of reactants and conversely, the effect of heat release on the turbulent mixing, are problems that have been studied for many years. There has, however, recently been a quickening of interest in, and attention to, this problem which is reflected in a recent review of turbulent reacting flows [14] ; the proceedings of a meeting largely devoted to turbulent reacting flows, Murthy [15] ; and a special issue of the journal, Combustion Science and Technology, devoted to turbulent reacting flows (to appear in 1976). While there are a variety of reasons for this increased interest, perhaps the most important is the need to control and, indeed, to reduce the harmful emissions from combustors in stationary and moving power plants. To achieve this a deeper understanding of the mixing and chemical processes is required than was needed to design such combustors to produce the requisite performance.

Despite an extensive literature, it must be said that our detailed knowledge of turbulent flows with chemical reactions is not satisfactory and that considerable additional work of both a theoretical and experimental nature must be carried out before there will be substantial improvement in that knowledge. Given the situation of great technical importance on the one hand and of great complexity on the other, it is not surprising that present research efforts involve a spectrum of approaches. Spalding and his co-workers (e.g. [16]) have incorporated chemical reactions into their various computer codes for turbulent flows of great complexity and have shown that the numerical techniques for the analyses of these flows do not present a major problem. With improved computational power expected over the next few years, it does not appear that the solution of large systems of partial differential equations, which hopefully describe the essential physics and chemistry of turbulent flows of practical interest, will offer a major obstacle. However, despite recent progress, the details of the physics of the turbulence and the interaction of the turbulence and chemistry as they appear in these equations certainly require considerable additional effort. To improve understanding of these phenomena some research workers have focussed attention on geometrically simple turbulent flows with simple composition and simple chemical mechanisms. This has been the direction of our own research.

The case of fast chemistry

Libby and Williams [14] discuss the various categories of turbulent reacting flows in terms of limiting cases; these give emphasis to particular aspects of a complex situation. In our own research we have focussed on one such limit, that of "fast chemistry". This limiting case has been the one attracting the greatest attention because its thorough treatment is a prerequisite for an attack on more complex, perhaps more practical flows. Although it is the best understood case of turbulent reacting flows at the present time, it is still not completely treated.

Consider the fast-chemistry limit in the case of non-premixed reactants. The chemical system is assumed to be represented by two reactants, generally thought to be fuel and oxidizer, and by a single product.* Figures 1a and 1b show two

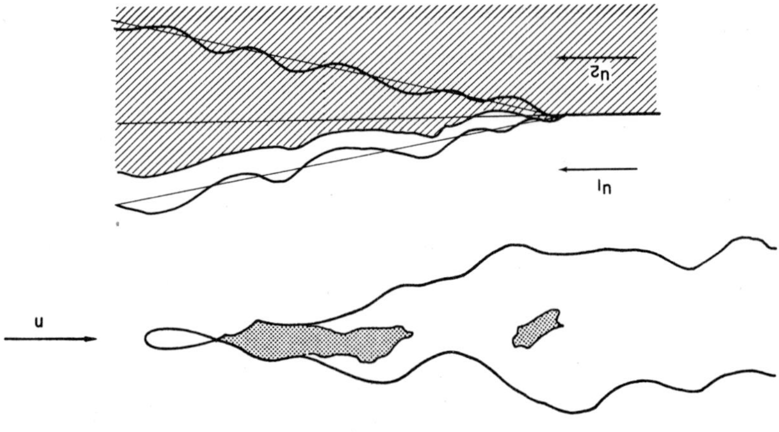

Fig. 1 Idealized flows involving non-premixed reactants and fast chemistry. *a.* The two-dimensional mixing layer. *b.* The fuel jet discharging into an oxidizing stream.

situations relevant to the present discussion. In Fig.1a we show a two-dimensional mixing layer with fuel in one stream and oxidizer in the other. In general the two streams have different velocities but the mean pressure is assumed constant throughout the flow field. If the initial boundary layers are turbulent but relatively thin, then under some circumstances the mean properties of this flow meet similarity requirements and are described by a single variable, $\eta \equiv x_2/x_1$. In more general cases we must consider that properties depend on x_1 and η. In Fig.1b we show an axisymmetric fuel jet discharging into a co-flowing oxidizer stream, again in general with different velocities in the two streams but with the mean pressure taken to be constant. In this case the mean flow properties depend on the cylindrical radius r and the streamwise distance x.

* Reference [14] points out the dangers of using global reaction mechanisms and rates in turbulent flows. Since this provides an extensive bibliography on the subject only the most pertinent references are included here.

The concept of fast chemistry has as its physical basis a high chemical kinetic rate at the molecular level; in fact, sufficiently high that the two reactants never coexist, i.e., one reactant is always present in sufficient amounts to consume all of the other reactant. Clearly, this assumption precludes consideration of some important aspects of turbulent reacting flows, e.g., those connected with ignition and extinction. Nevertheless there are situations of practical interest involving temperatures, pressures, physical scales, and velocities, where the fast-chemistry assumption is reasonable. In addition there are flows where the principal energetic reaction can be idealized by the fast-chemistry model while secondary reactions (for example, those producing pollutants) are determined by finite kinetics.

To be more precise about the fast-chemistry assumption consider the chemical reaction

$$M_1 + M_2 \rightarrow M_3 \quad . \tag{4}$$

The fast-chemistry concept may be stated

$$Y_1(\mathbf{x},t)\; Y_2(\mathbf{x},t) = 0 \tag{5}$$

where Y_i denotes the mass fraction of species i.

Figures 1a and 1b show the customary interface between the interior turbulent flow and the external irrotational flows. The presence of these contorted interfaces in turbulent shear flows leads to the phenomenon of intermittency which, we shall see later, plays an important role in the chemical behaviour. The figures also show another surface embedded within the turbulent fluid; this is the flame sheet which is a result of the non-coexistence of reactants indicated by equation (5). The flame sheet separates fluid with one reactant, product and inert diluent from fluid with the other reactant, product and diluent. Product is formed by chemical reaction at the molecular level at this surface. We shall discuss later how the statistical geometry of a flame sheet can be studied at least in one limiting case. For the present we note simply that, following current views on the behaiour of the irrotational-turbulent interface, a large number of overhangs (which can lead to multiply-connected regions) have not been included in Figs 1a and 1b. On Fig.1b, however, is shown schematically an element of encapsulated fuel which has been torn from the main portion of fuel-rich fluid, and carried downstream to be eventually consumed.

In most practical situations where the fast-chemistry limit is approached, the chemical reaction leads to sufficient heat release for there to be significant density variations; thus an interaction between the fluid mechanical and chemical processes occur. It would be expected therefore that for the flows of Fig.1 the spreading rate of the mixing layer and of the fuel jet and the rate of decay of the latter would be altered by the chemical reaction and heat release. As Libby and Williams [14] note, the interaction between heat release and turbulence is so poorly understood that there is controversy regarding such a basic point as whether heat release increases or decreases turbulent kinetic energy. Bray [17] shows that competing effects are involved which might account for the controversy: turbulent kinetic energy is generated by shear associated with the heat release but the heat release leads to a dilatation which destroys that energy. Given this situation, which of course needs quantifying, the controversy is perhaps not surprising.*

A much simpler case arises when the reactants are highly diluted in a background gas which is isothermal; for then there is no interaction between the chemical and the mixing processes. This is the case which has attracted the most research attention because of its fundamental position and relative analytic simplicity. This model appears to have been considered first by Hawthorne *et al.* [18] and later by Toor [19]. More recently, following the intensification of interest in turbulent reacting flows, there has been a series of studies of this case reported by Lin and O'Brien [20], Bilger and Kent [2], Libby [22-23], Alber and Batt [24] and others. Despite the efforts represented by this literature no completely effective approach to the relatively simple flows suggested by Figs 1a and 1b exists.

One of the great advantages to studying the case of highly dilute reactants undergoing fast chemical reactions is that experimental data on passive scalars, such as temperature or the concentration of a trace species, are directly applicable. Thus, for example, the statistical geometry of the flame sheet can be inferred from the related behaviour of a contour of either constant temperature, or constant concentration and the fluid mechanical properties at such a contour can be studied by the highly developed techniques of conditioned sampling.

2. HOT-WIRE ANEMOMETRY FOR HELIUM-AIR MIXTURES

2.1 *Preliminary Remarks*

In our first paper on our hot-wire work in helium-air mixtures (Way and Libby [25]) the motivation for this effort is stated:

> "There are many turbulent flows of practical interest involving significant density fluctuations, e.g., in the wake of bodies in hypersonic flight, in propulsion units employing supersonic combustion, and in boundary layers with external streams having supersonic or hypersonic velocities. In many of these flows measurements of mean quantities such as velocity, temperature, and composition have been made and have been incorporated in methods of analysis. However, few measurements of fluctuating quantities in such flows have been made because of the experimental difficulties involved. In fact, Laufer† recently stated 'It is somewhat disconcerting, for instance, that since the work of Kistler** ... ... no experiments have been reported on turbulent fluctuations in a compressible flowfield above Mach 4.' "

---

* Professor K.N.C. Bray and the author have recently completed and submitted for publication a study involving application of the Bray-Moss theory for premixed reactants to the planar turbulent flame. The turbulent flame speed and the alteration of the turbulent kinetic energy by the heat release are calculated as part of the solution. The results clarify the interaction between turbulence and heat release and the competition between dilatation and shear, and fully support the Bray hypothesis.

† Laufer, J. (1969). *In* "Compressible Turbulent Boundary Layers" (M.H. Bertram, ed). NASA SP-216, pp.1-13.

** Kistler, A.L. "Fluctuation Measurements in a Supersonic Turbulent Boundary Layer", *Phys. Fluids* 2, 290-296, 1959.

Five years later the paucity of experimental data to which the above remarks refer still remains.

The notion of using hot-wire anemometry for measuring quantities other than velocity is not new. Corrsin [26] set forth the basic ideas of measuring temperature and concentration of a second species by extended hot-wire techniques in 1949. Nevertheless, while the measurement of temperature simultaneously with velocity is relatively commonplace (see, for example, Bradshaw [6]) direct application of Corrsin's ideas to measuring concentration is uncommon. In the following discussion for simplicity and clarity attention is focussed on the specific problem of measuring the streamwise velocity component, denoted u, and of the mass fraction of helium, c. The extension of the method to the measurement of a second velocity component has been carried out and reported by Stanford and Libby [27] who added a swept sensor. The determination of the normal velocity component depends on knowledge of the corresponding values of u and c. Information about u, v, and c (or u_1, u_2, c in subscript notation) adds greatly to the utility of the experimental results but the experiments are, of course, more complex.

The straightforward approach to the measurement of concentration along the lines suggested by Corrsin can be understood as follows. Consider a low-speed, isothermal flow involving fluctuations in velocity and helium concentration. If two hot-wires of different characteristics, i.e. different operating temperatures, diameters, etc., are inserted in such a flow and operated in a constant temperature mode by suitable electronic equipment, and if their outputs, E_1 and E_2, are recorded, then roughly we can use King's law for each wire and each concentration of helium and consider

$$\begin{aligned} E_1^2 &= A_1(c) + B_1(c)\,u^n \\ E_2^2 &= A_2(c) + B_2(c)\,u^n \end{aligned} \tag{6}$$

where $n \approx 0.5$, where $A_i(c)$ and $B_i(c)$, $i = 1, 2$ are obtained by calibration in various mixtures with known mass fraction c of one of the two species, and where u is the streamwise velocity assumed considerably larger than any velocity component normal thereto. Thus in principle, measurement of a voltage pair, E_1 and E_2, in a turbulent flow can be inverted via equations (6) to obtain values of u and c.

2.2 *An Interfering Probe*

The above approach, at least for helium-air mixtures, does not appear to be practical. The basic difficulty is that increases in c and u lead to increases in voltage for each sensor. Contrast this with the desirable situation of having one sensor yield the concentration and the other velocity. In common with other workers, we were never able to obtain sufficient difference in the operating characteristics of two hot-wires to avoid a severe problem of ambiguity; thus, we could not, for a given voltage pair, E_1 and E_2, distinguish between a high velocity in a low concentration of helium from a lower velocity in a higher concentration of helium. This situation is reported by Way and Libby [28].

The way out of the difficulty is a probe which involves two sensors, one a hot film whose power requirements are described by a King's law relationship (i.e. like equation (6)), and a hot-wire close to the film. "Close to" in the present context means in the thermal field of the film. We show schematically such a configuration in Fig.2. If the flow conditions and spacing between the film and wire

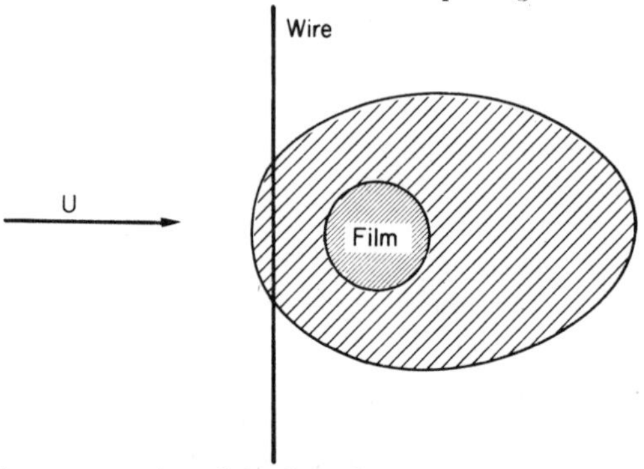

Fig. 2 A schematic representation of the flow about one interfering probe.

are such that all of the energy needed to keep the wire at its operating temperature comes from electrical heating, then the wire also follows a King's law and the ambiguity problem discussed previously persists. However, if part of the energy is in the form of thermal input from the film and the remainder from electrical heating, then the wire does not follow a King's law. Consider for example a situation in which the velocity of a stream is held constant while the concentration is increased. Due to the high thermal conductivity of helium the electrical power to the film must be increased to hold its temperature constant. However, as the concentration of helium increases, the thermal field of the film expands; it is thus not clear whether more or less electrical heating would be needed to keep the wire at an average constant temperature. We thus have the possibility of adjusting the spacing of the wire from the film and the operating temperatures of the wire and the film so that under the situation described previously, the wire voltage at least does not go up, perhaps stays constant and may even decrease. In this last case the thermal heating from the film increases to such an extent that the electrical power requirement of the wire actually decreases.

The practical realization of this possibility is shown in Figs 3a and 3b. Shown in both are so-called calibration maps of two different, two-sensor probes. These maps represent plots of wire voltage squared, E_w^2, versus film voltage squared, E_f^2, with contours of constant concentration of helium and constant x-wise velocity indicated. In Fig.3a the velocity contours are given in cm/sec whereas in Fig.3b they are in m/sec. Both probes exhibit the general features indicated as being desirable earlier; we see that for a fixed velocity, increases in the concentration of helium lead as expected to increases in the voltage to the film. However, the wire receives an increasing amount of its energy from thermal heating from the film and its voltage decreases. These characteristics remove the ambiguity referred to previously.

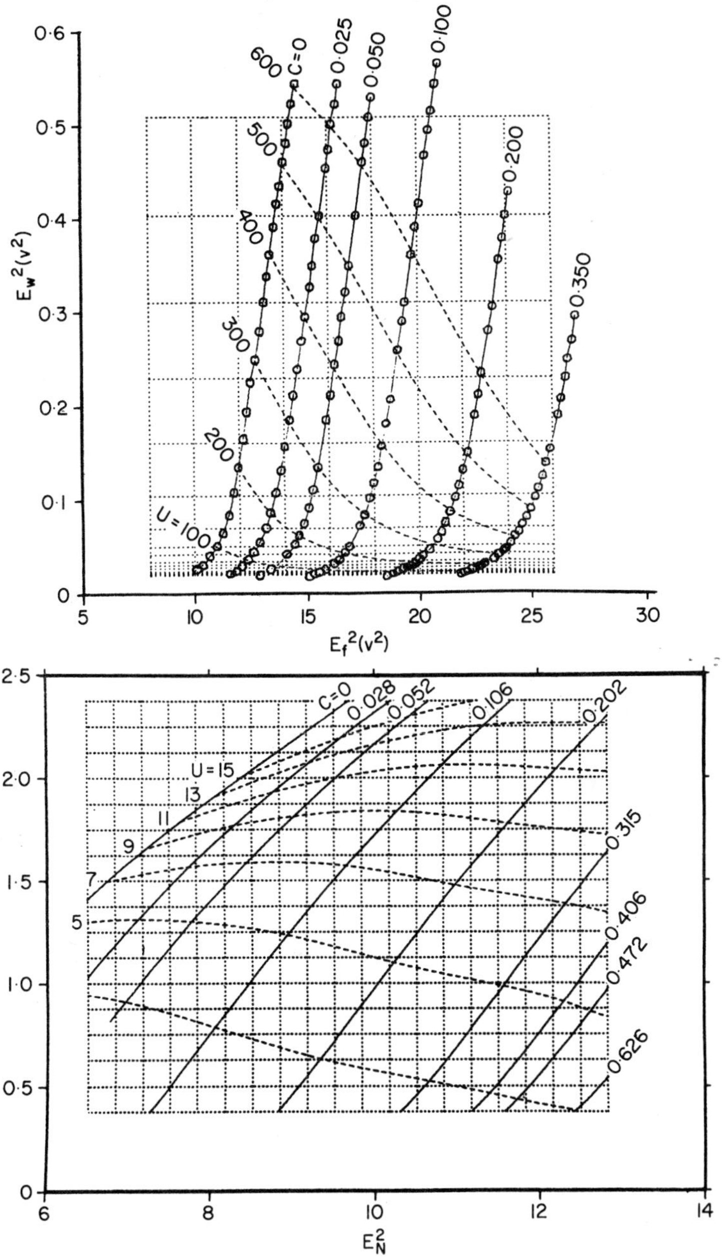

Fig.3 **Calibration maps.** ***a.*** **From Way-Libby [25].** ***b.*** **From Stanford-Libby [27].**

A little reflection indicates that the operating conditions and geometry of the film and wire must be arranged to avoid, over the range of velocities and concentrations of interest, two undesirable limiting operating conditions. If the interference between the sensors is insufficient, as tends to be the case at high velocities in low helium concentrations, the wire and film behave independently, each following a King's law; the ambiguity problem described earlier then precludes satisfactory operation. If the interference is excessive, all of the energy necessary to maintain the wire at the specified average temperature will be obtained from thermal heating and no voltage signal will be obtained. This tends to be the situation at low velocities in high concentrations of helium. From the point of view of probe operation it is preferable to have under investigations a flow which has low velocities in air and higher velocities in helium; in this case the turbulent data tends to be in an area running from the lower left to the upper right in Figs 3a and 3b. A two-dimensional or circular jet of helium discharging into a slower-moving airstream is a flow which leads to this behaviour; we are setting up such an experiment at present.

2.3 *Probe Calibration and Data Reduction*

In using a probe of the type described above one needs to know beforehand (as in all hot-wire anemometry) the range of expected operating conditions: in the present instance, the range of velocities and concentrations. The concentration range is then divided into six to nine concentrations; mixtures of known concentration of helium and air covering this range are then prepared. The probe is calibrated in a calibration jet, much as is a conventional single wire used in air, but in this case in each of the mixtures.

To appreciate the techniques involved in utilizing these probes it is important to distinguish between calibration data and turbulent data. The former are data involving measurements of the voltages, E_f and E_w, corresponding to known u–c pairs and resulting in the calibration maps shown in Fig.3. Turbulent data are obtained when the probe is in a turbulent flow with fluctuating velocities and concentrations; in this case the measured voltage pair, E_f and E_w, are to be used to yield by means of an appropriate data reduction procedure the desired u–c pair.

The inversion of a voltage pair to a u–c pair has been carried out in several ways. In references [25] and [27] a rectangular mesh was overlaid on the contours of constant concentration and velocity and by spline fitting, the calibration data were transferred to the grid points of the mesh. This is the grid shown in Figs 3a and 3b. A voltage pair, E_f and E_w, corresponding to turbulent flow will lie within one grid rectangle. The related u–c pair is obtained from a two-variable interpolation involving the surrounding nine grid points as indicated in Fig.4.

When the voltage pair corresponds to either a small helium concentration or to one close to pure helium, the interpolation to obtain the correct concentration will involve mesh points beyond the $c = 0, 1$ contours and thus to values of $c < 0$ and $c > 1$, i.e., calibration data must be extrapolated beyond physically significant values. Thus, notions of concentrations with negative and greater than unity values arise.

A second data reduction scheme is based on the observation that the contours of constant concentration in the $E_f^2 - E_w^2$ variables are smooth and well-behaved.

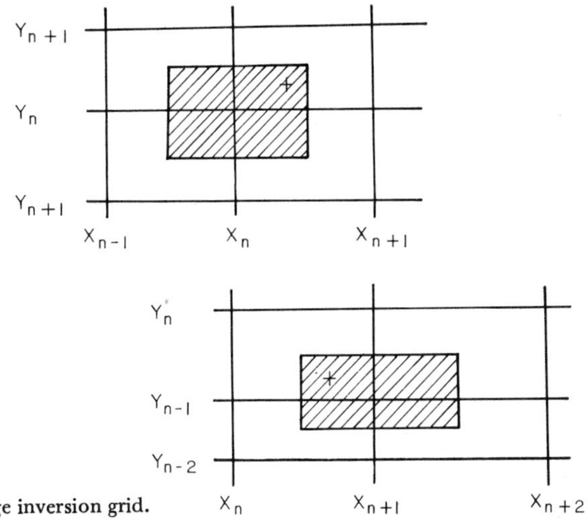

Fig.4 The voltage inversion grid.

A representation

$$c = a_1(X)Y + a_2(X)Y^2 + \ . \ . \ . \tag{7}$$

is suggested where $X = E_f^2$, $Y = E_w^2$, and where the $a_i(X)$ coefficients are in turn polynomial functions of the indicated variable. Thus the film and wire voltages give explicitly the concentration. The film behaves by itself much as a conventional hot-wire with calibration coefficients discretized functions of concentration. With the concentration known, a King's law in the form

$$u = \left[\left\{X - A(c)\right\} / B(c)\right]^{\frac{1}{n}} \tag{8}$$

gives explicitly the corresponding u-velocity component.*

We have not had sufficient experience with these two data-reduction methods to choose one over the other. The second has the advantage that extrapolation beyond the physical boundaries of $c = 0, 1$ is automatically handled. The need for such extrapolation may be seen as follows. Suppose that a pair of voltages yielding X and Y are collected in the course of an experiment and in the data reduction are substituted into equation (7). Suppose further that the true concentration of helium is zero, but that due to experimental errors the apparent concentration is negative. Equation (7) will simply give a negative value for c. We shall discuss later how we use such a physically unacceptable value in the assessment of accuracy. From the point of view of accuracy and computing speed it appears likely that the former system is preferable.

---

* Dr. John LaRue has recently improved Eq. (8) by observing that for constant temperature operation a King's law with more physical content is $X = \lambda(c)(A(c) + B(c)(\rho u)^n)$ where $\lambda(c)$ is the thermal conductivity of the mixture. In this case $A(c)$ and $B(c)$ are nearly constants and the data reduction scheme placed, at least in part, on a firmer basis.

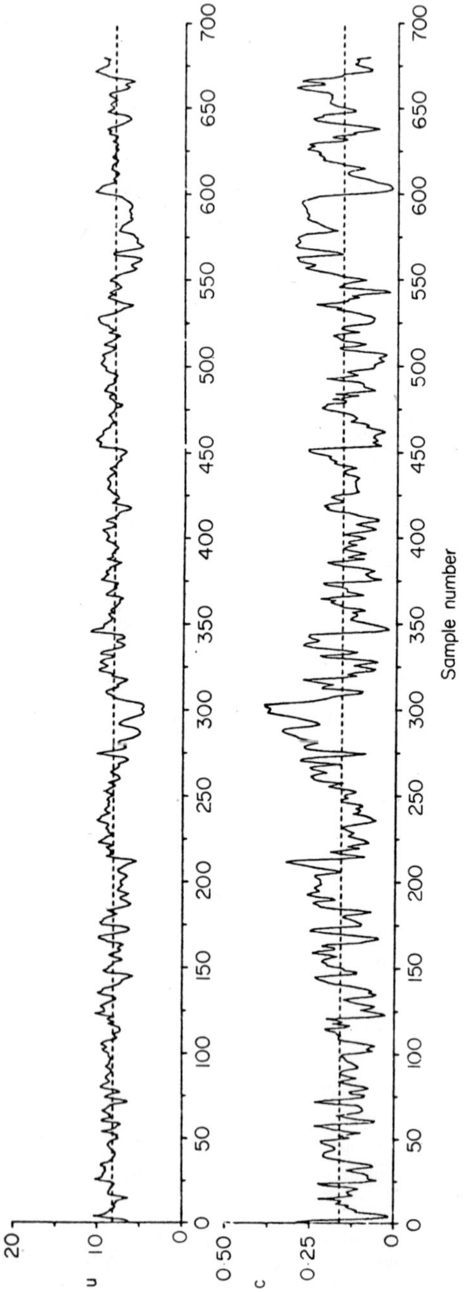

Fig.5 **Typical time series in the *u*-velocity component and helium concentration.**

Because of limitations of the digital tape recorder available to us, we record between two and four thousand samples per second per channel depending on the number of channels being recorded. Typically recording continues long enough for 250,000 samples to be collected. This is generally sufficient for adequate statistical processing even with severe conditioning, i.e., even when some conditioning criterion requires exclusion of a large percentage of the data.

One question of interest is the computing time needed to invert one voltage pair to the corresponding $u-c$ pair. All the data reduced so far have used a CDC 3600 computer, a relatively old and slow machine. The original Way-Libby results took six seconds for each inversion; refinements have reduced the inversion time to 1.4 seconds/second of data, i.e., to nearly real time data reduction.

The results of the inversion procedure are time series in the u-velocity component and the helium concentration. These time series can be employed to present results in a variety of forms as we shall discuss in more detail later. It is interesting to plot by computer some of the time series since such a primitive display of results is physically appealing. Figure 5 shows the time trace of u and c for a total time period of approximately one third of a second; this might be considered a simulation of an oscilloscope display of probe output with data reduction being by way of a dedicated on-line computer.

From the foregoing description it will be seen that a combination of laboratory and computing effort is associated with this extended hot-wire anemometry. In terms of the time involved in performing and completing one experiment, we find that the data collection time is considerably less than the data-reduction and data-interpretation times. This is common for modern experimental turbulence research involving digital techniques. In our particular studies the data reduction effort is probably considerably greater than that involved in any other current turbulence research because the inversion of the electrical signals to fluid mechanical quantities involves complex functional relations in contrast to the simple, frequently linearized relations in the usual hot-wire and cold-wire techniques. However, the interpretation effort based on time-series analysis is comparable with that carried out in many other investigations.

In Fig.6 we show a block diagram of the electronic equipment needed to use the two-sensor probe. The signals from two constant-temperature anemometer bridges are fed to instrumentation identified as "Analogue Processing" which performs several functions; it scales and offsets the bridge signals to take advantage of the full range of the downstream analogue-to-digital converter (ADC). In addition the bridge signals are taped off, fed into square circuits and thus onto a storage oscilloscope which permits the calibration map such as shown on Fig.3 to be displayed during a test. The down-stream equipment culminating in the digital tape recorder is standard and is operated in a standard fashion.

Figure 6 also indicates the pressure-measuring equipment. This permits the calibration velocities to be determined from the plenum pressure which is recorded as part of the calibration data.

We intend shortly to carry out an experiment in which the signals will be recorded on both digital and analog tape recorders. The analog signals will be subsequently digitized; this is a standard procedure in experimental turbulence and

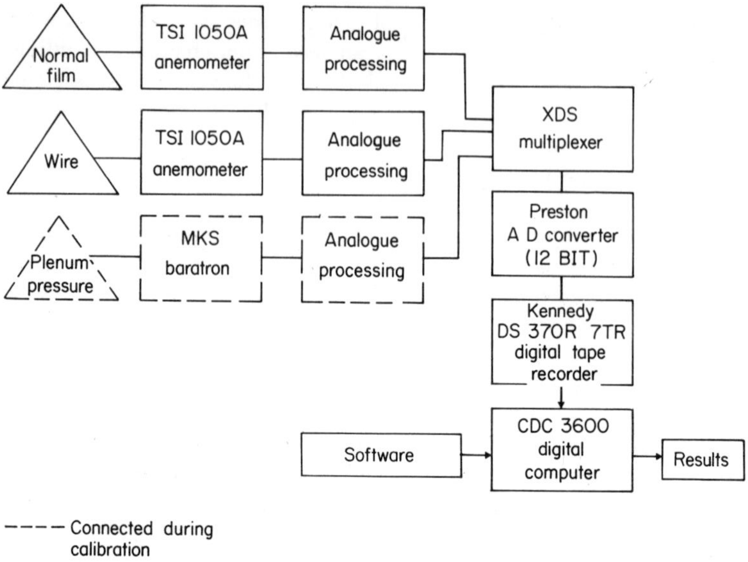

Fig. 6 The block diagram of electronic equipment.

has significant advantages in data analysis. It is generally less accurate, however, and accordingly, we have hitherto preferred to record directly in digital form. The experiment will enable us to assess whether the analogue technique is well-suited to our particular anemometry.

2.4 *Assessment of Accuracy*

Consideration of the potential errors in the calibration, data-collection and data-reduction phases of obtaining measurements in turbulent helium-air mixtures shows clearly the need to assess the self-consistency and accuracy of the system during each experiment. We describe here the several procedures we have followed.

The most obvious check in assessing accuracy of the data-reduction system is to treat calibration data as turbulent data and to compare predicted and known values of velocity and concentration. This procedure frequently exposes short-comings in the extrapolation and curve-fitting schemes and is a crucial, first step in validating the procedure. When operating correctly our data-reduction programs recover the calibration values within 0.001 in mass fraction for low concentrations and within 1% for the velocity and for the higher concentrations.

Another check on self-consistency can be carried out when the turbulent data correspond to a turbulent flow of known and constant concentration of helium. This can be established in several ways depending on the flow configuration under consideration. For example, for the tests described in [25] involving the discharge of a helium jet into quiescent air, we put a mailing tube over the orifice and passed through it air containing 10% helium by mass. The accuracy to which the actual concentration is achieved experimentally provides a measure of overall system performance. In all of our tests in which this check has been used, the mean concentration is accurate within 0.001 with an rms error of about the same magnitude.

These same test data (those corresponding to a known concentration of helium in a turbulent flow) can be used in a different fashion. If the data-reduction system is modified so that the known constant concentration is imposed, the output of the two sensors provides two measures of velocity. A comparison of the two apparent velocities gives another measure of system accuracy. In [25] the two mean velocities and rms fluctuations agreed within three percent.

Finally, we mention that there is no override criterion in the data-reduction program to prevent negative helium concentrations being indicated. The frequency with which voltage pairs, upon inversion to $u-c$ pairs, give negative helium concentrations itself provides a measure of system accuracy. The number of data points in a time series with negative values for c depends, of course, on the mean concentration and on the intensity of the concentration fluctuations. Indicative results are: for $\bar{c} = 0.014$ and $\overline{c'^2} = 0.0133$ we find that c is negative for one sample in about 400.

Finally it is mentioned that care is taken to assure that the temperature of both the air and the helium remains close (within 0.5 deg C of the temperature of the gas mixtures used in calibration). We do so because temperature heterogeneities can clearly introduce errors.

2.5 *Applications to Some Variable Density Shear Flows*

a. Density fluctuations. The techniques described above have now been applied to several flow situations. The original Way-Libby work involved a circular jet of helium discharging into quiescent air. Stanford and Libby carried out experiments in a porous tube with turbulent air introduced through one open end and with pure helium injected uniformly through the porous cylindrical surface. Since the latter set of data is more extensive it is chosen to illustrate the sort of conventional mean and intensity data obtained. Data are identified by their spatial location in terms of the ratios x/L and r/R where x is the axial coordinate, L the length of the pipe, r the radial coordinate and R the radius of the pipe.

In Fig.7 we present distributions of several flow variables at three axial stations in the pipe. In Fig.7a we show the distributions of axial velocity normalized with respect to the velocity on the axis at the first measuring station, $x/L = 0.5$. We see that the effect of injection is to accelerate the flow along the axis of the pipe. The effect of the low density of the injected fluid is to accelerate somewhat the mixture off-axis. Figure 7b shows the build-up of the helium concentration as the flow evolves. Figures 7c and 7d show the considerable increase in the intensity of the velocity and concentration fluctuations as the porous surface is approached. One of the short-comings of these data, which we shall remedy in the future, is that they do not provide adequate coverage of the near-wall region where radial gradients of flow properties are steepest.

The analysis of the time series which is generated by our data-reduction procedure involves the well-known methods of time-series analysis and Fast Fourier Transform. Gibson [7] provides a valuable review of, and bibliography for, those aspects of these methods of greatest value in experimental turbulence research. Basic results from their application are the mean values and the statistics of the fluctuations of the several variables both individually (as shown in Fig.7) and in

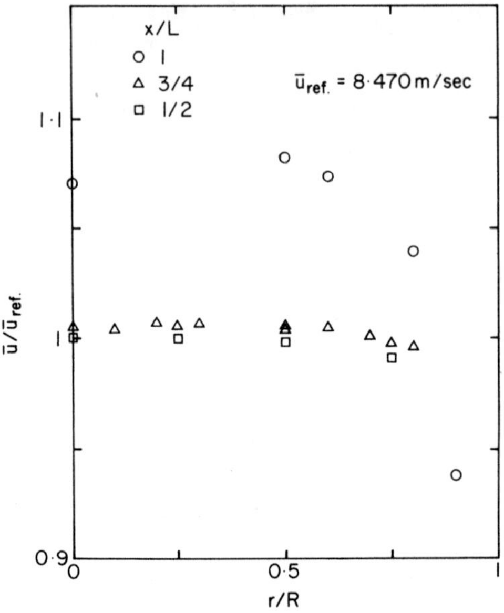

Fig. 7a The distribution of mean values and intensities in a porous pipe. The mean u-velocity component, $\bar{u}/\bar{u}_{ref}$.

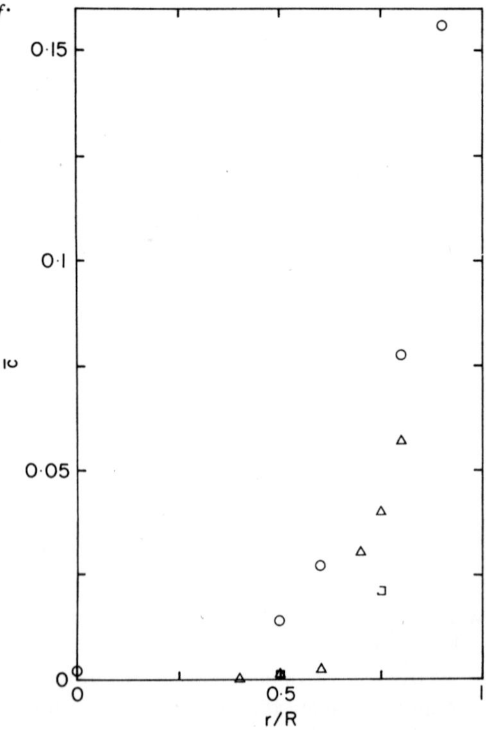

Fig. 7b The distribution of mean values and intensities in a porous pipe. The mean concentration, $\bar{c}$.

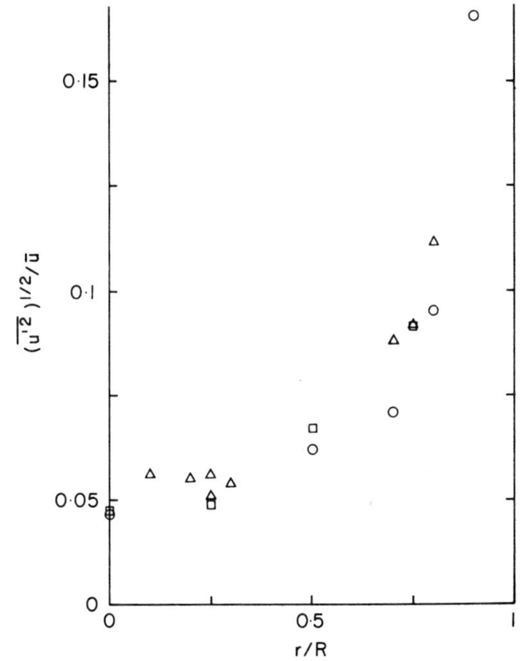

Fig. 7c The distributions of mean values and intensities in a porous pipe. The relative velocity intensity, $(\overline{u'^2})^{1/2}/\bar{u}$.

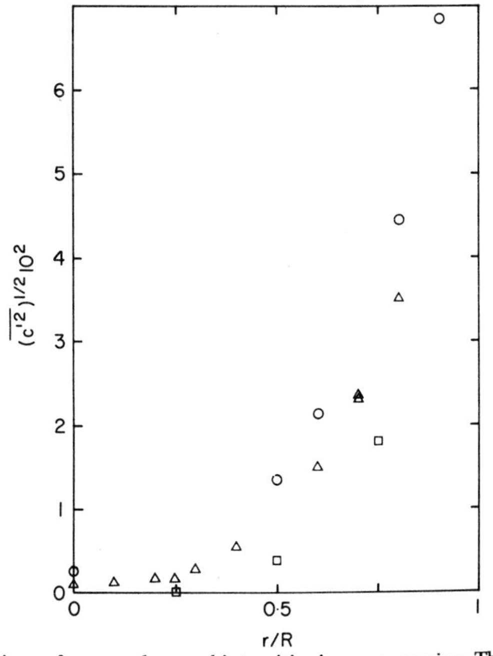

Fig. 7d The distributions of mean values and intensities in a porous pipe. The concentration intensity, $(\overline{c'^2})^{1/2}$.

cross-correlations. In addition the use of digital techniques offers several other methods for displaying data some of which are particularly illuminating; we shall discuss these in the following sections. Since the variability of the density provides the principal motivation for our research, density fluctuations are considered first.

With the assumption for low-speed, isothermal flows that density fluctuations are due solely to concentration fluctuations, the time series for c can be converted to a new time sereis in density. The linking equation is

$$\frac{\rho}{\rho_o} = \frac{1}{(1 + wc)} \tag{9}$$

where ρ_o is the density associated with the fixed pressure and temperature for $c = 0$; W_a, W_c are the molecular weights of the two species, air and helium, and $w \equiv (W_a/W_c - 1) = 6.22$. Evidently from Eq.(9), the determination of the mean and fluctuations of density involves a non-linear operation on the concentration.

To give some sense of the relation $\rho = \rho(c)$, consider an approximation for the intensity, $\overline{\rho'^2}/\bar{\rho}^2$, a quantity which serves as a measure of the extent of the density variations in a given flow. The tempting approximation, $wc << 1$, applied to Eq.(9) is of only limited applicability because of the large value of w. However, if the fluctuations in concentration are assumed small (more precisely, if $wc'/(1 + w\bar{c}) << 1$), then we obtain the approximation

$$\frac{(\overline{\rho'^2})^{1/2}}{\bar{\rho}} = w\bar{c}/(1 + w\bar{c}) \quad (\overline{c'^2})^{1/2}/\bar{c} \tag{10}$$

which indicates that the coefficient linking the relative density intensity $(\overline{\rho'^2})^{1/2}/\rho$ to the corresponding concentration intensity depends upon the mean concen tration levels. For small mean concentrations (i.e. $w\bar{c} << 1$) this factor is approximately $w\bar{c}$. When this condition is satisfied, there is no significant dynamic coupling via the density between the velocity and concentration; the helium is then dispersed as a passive scalar. At the other extreme, $w\bar{c} >> 1$, the multiplicative factor approaches unity: this would arise only when $\bar{c}$ is nearly unity for helium-air mixtures.

Figure 8 compares the results we have obtained from several experiments with the estimate given by Eq.(10). Reasonable agreement prevails. We have excluded from consideration in Fig.8 data with intense density fluctuations since they violate the inequality leading to Eq.(10). We refer to cases with a relative density intensity as high as 60%.

b. Favre averaging. The time series in the density can be used to compute the mass-averaged mean and fluctuating quantities, i.e., the experimentally determined quantities after Favre. A comparison of the two means for averaging is of interest since as indicated earlier there appears to be no agreement on which of the two approaches is most appropriate for the treatment of turbulent flows with variable densities.

We have already indicated the simplicity of the describing conservation equations afforded by Favre averaging. That this simplicity has *not* led to wide accept-

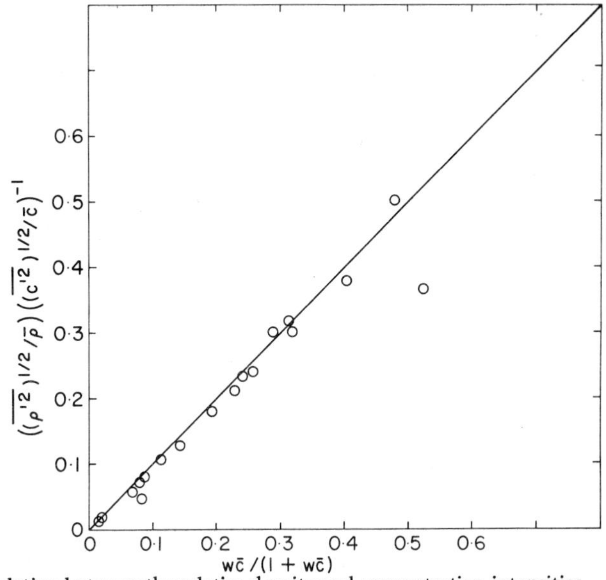

Fig.8 **The relation between the relative density and concentration intensities.**

ance of Favre averaging appears to be due to suspicion that the simplicity is deceptive and that special modelling to effect closure is required. To see the nature of the problem consider the momentum equation for boundary layer flows according to the two means of averaging; we have without approximation,

$$\frac{\partial}{\partial x_k}(\bar{\rho}\bar{u}_k\bar{u}_1) = -\frac{\partial \bar{p}}{\partial x_1} - \frac{\partial}{\partial x_2}(\bar{\rho}\overline{u'_1u'_2} + \bar{u}_1\overline{\rho'u'_2} + \bar{u}_2\overline{\rho'u'_1} + \overline{\rho'u'_1u'_2}) \qquad (11)$$

$$\frac{\partial}{\partial x_k}(\bar{\rho}\tilde{u}_k\tilde{u}_1) = -\frac{\partial \bar{p}}{\partial x_1} - \frac{\partial}{\partial x_2}(\overline{\rho u''_1u''_2}) \qquad (12)$$

Comparison of these two equations indicates that the extra terms on the right side of Eq.(11) must be hidden in the Favre-averaged quantities on the left side of Eq.(12). This can be seen as follows: We have already shown (cf. Eq.(3)) that

$$\tilde{u}_k = \bar{u}_k + \overline{\rho'u'_k}/\bar{\rho} \quad .$$

In addition the definitions of the double-primed fluctuations and introduction of $\rho = \bar{\rho} + \rho'$ readily lead to the result

$$\overline{\rho u''_1u''_2} = \bar{\rho}\overline{u'_1u'_2} - \overline{\rho'u'_1}\;\overline{\rho'u'_2}/\bar{\rho} + \overline{\rho'u'_1u'_2} \quad .$$

Substitution into Eq.(12) shows that the mass transfer and triple correlation terms on the right side of the conventionally averaged equation are automatically contained in the $\tilde{u}_1$ and $\tilde{u}_2$ components. This suggests that a difference in modelling may in fact be required according to which of the two means of averaging is adopted.

We have used our helium-air data to compare $\tilde{u}_1$ with $\bar{u}_1$ and $\overline{\rho u_1'' u_2''}$ with the usual Reynolds stress term $\bar{\rho}\,\overline{u_1' u_2'}$. The accuracy of $\tilde{u}_2$ and $\bar{u}_2$ does not warrant their comparison but of course the fluctuations, u_2'' and u_2', can be considered in terms of intensities and correlations. The results of this comparison for a wide variety of flow conditions indicates that as expected $((\tilde{u}_1/\bar{u}_1) - 1)$ increases with the relative density intensity, $(\overline{\rho'^2})^{1/2}/\bar{\rho}$, reaching values of 5-6% when that intensity is 60%. These relatively small differences indicate that, as suggested earlier, high accuracy in experiments may be needed to determine the differences in some measured quantities averaged according to the two means.

In contrast the ratio $(\overline{\rho u_1'' u_2''}/\bar{\rho}\,\overline{u_1' u_2'})$ can become considerably greater than unity when the mass transfer term $\bar{u}_1 \overline{\rho' u_2'}$ is important. Stanford and Libby [27] report values up to 1.85 for the ratio $(\bar{u}_1 \overline{\rho' u_2'}/\bar{\rho}\,\overline{u_1' u_2'})$ which provides a measure of the relative size of the principal mass-transfer term to the usual Reynolds stress term; we have found in other experiments, as yet unpublished, values of this ratio of roughly −8!. Both the sign and magnitude of these ratios are of interest because the signs of $\overline{\rho' u_2'}$ in both experiments imply a counterflux of helium concentration, i.e., at net flux toward regions of higher helium concentration. The explanation for this result appears to reside in a weak transverse pressure gradient in both flows providing the data under consideration. These gradients, inferred from measured intensities $\overline{u_2'^2}/\bar{u}_1^2$, are of the size known to exist (and to be unimportant) in many turbulent shear flows. However, when density fluctuations are present, these weak transverse gradients appear to have a significant effect on transverse momentum exchange and to account for the sign of the $\overline{\rho' u_2'}$ terms. Assessment of the magnitude of these terms requires further study. The implication of this result is that neglect of the mass transfer term $\bar{u}_1\,\overline{\rho' u_2'}$ compared to the Reynolds stress term when conventional averaging is used in variable density turbulent flows may be greatly in error if small transverse pressure gradients exist, and that any agreement between prediction and experiment in variable density flows with such gradients and with only the $\overline{u_1' u_2'}$ term taken into account may be fortuitous.

We have already indicated that in the case of stratified flows the production of turbulent kinetic energy by body forces disappears upon Favre-averaging; the contribution of this agency must be hidden in the flux terms, and would have to be incorporated in the modelling of those terms for the resultant equation to display the observed sensitivity to body forces. In this connection, it should be noted that the modelling of body-force influences in variable-density flows is not well established even for conventional averaging. It now appears that the effect of weak transverse pressure gradients must also be taken into account. Thus the question of closure does not appear to provide grounds for preferring conventional over Favre averaging; our point here is that the modelling for both averaging methods requires more attention when density variations are significant. It does appear, however, that advantage should be taken of the simplicity afforded by Favre averaging since the modelling may be more direct and easier to rationalize.

It is interesting to note that Favre [29] has recently compared the two methods of averaging for flow situations involving various sources for variable-density effects. He has, however, not included in his considerations the case of interest here, that is, the mixing of gases of widely different molecular weights. Favre deduces that in boundary layer flows quantities involving the transverse velocity should differ more significantly than those involving only the streamwise velocity component. Our helium-air data support this deduction but suggest that the large values of $\overline{\rho' u_2'}$ in our experiments arise from weak normal pressure gradients.

c. Probability density functions. The mean values and intensities such as considered so far are typical of the results obtained from earlier experimental techniques based on analog methods. Digital techniques, however, allow a far more detailed statistical analysis usually expressed as probability density functions. Consider, as a random function, the u-velocity at a given space point **x**. Associated with this variable is a function $P(u;\mathbf{x})$ from which the usual mean values and intensities of the fluctuations can be computed according to

$$
\begin{aligned}
1 &= \int P(u;\mathbf{x})\, du \\
\bar{u} &= \int P(u;\mathbf{x})\, u\, du \\
\overline{u'^2} &= \int P(u;\mathbf{x})\, u^2\, du - \bar{u}^2
\end{aligned}
\tag{13}
$$

In fact all moments of the fluctuations of the u-variable can be obtained by obvious extensions of Eqs (13). The function $P(u,\mathbf{x})$ is called the probability density function (pdf) for the u-variable; corresponding pdf's exist for each space point and for each dynamic variable.

We have so far been indefinite about the integration ranges involved in the integrals shown. Because of the pressure fluctuations associated with turbulence, there are no sharp bounds on the pdf's involving the velocity but this is not the case with other variables. For example, the mass fraction of any species is bounded between zero and unity; thus the limits on the integrals associated with $P(c;\mathbf{x})$ are 0,1. Similar bounds can sometimes be identified for other variables; for example, if pressure and radiative effects can be neglected, it may be possible in some flows to identify a minimum temperature so that a definite lower limit on the integrals of $P(T;\mathbf{x})$ prevails. This feature of boundedness can cause fundamental difficulties with analyses, explicitly or implicitly based on the assumption of normal or nearly normal behaviour of the variables involved.

In developing a probability density function from a time series it is necessary to divide the range encompassing the variable into a certain number of "bins". The time series is then scanned, the various terms deposited in the appropriate bins, and the population in each bin counted. When all entries have been disposed of, the populations are then divided by the total number to give a percent in each bin. The percent values in each bin are "smoothed" in order to yield a pdf. This description of the probability density function also indicates its physical signifi-

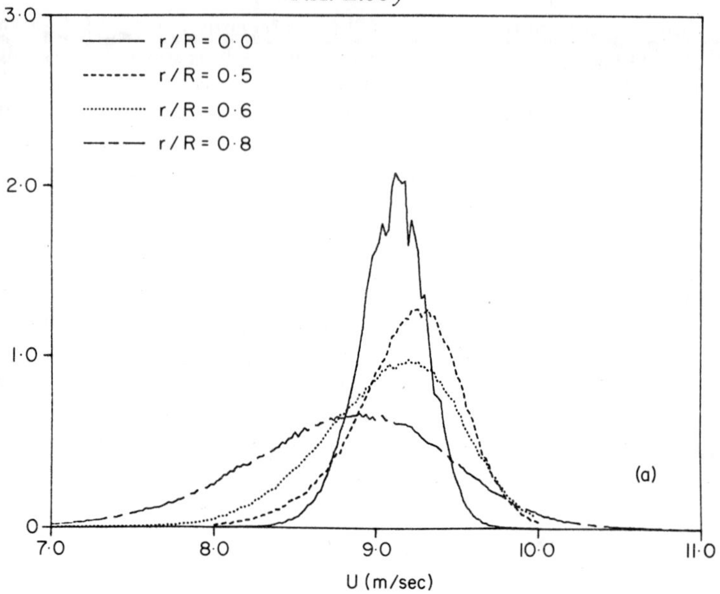

Fig. 9a The probability density functions in a porous pipe: $x/L = 1$. The u-velocity component, $x/L = 1$.

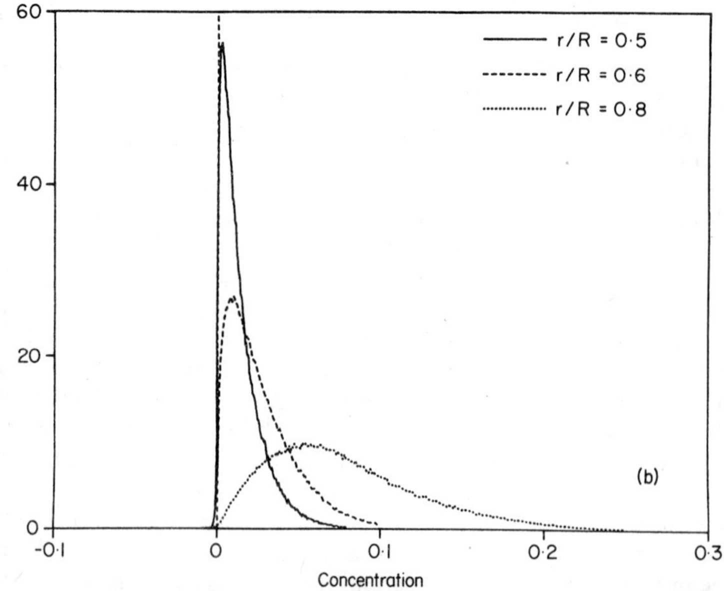

Fig. 9b The probability density functions in a porous pipe: $x/L = 1$. The concentration $x/L = 1$.

cance; at each space point each fluid mechanical variable has a pdf which implies the infinitesimal percent of the time the variable in question is within an infinitesimal range of values in the neighborhood of the specified value.

From a practical point of view the selection of the bin size involves a compromise. As the bin size is made smaller, the number of entries gets smaller and with a finite (though generally large) number of total entries available, the pdf becomes irregular. On the other hand an excessively large bin size provides a poorly resolved pdf.

In Fig.9 we show several pdf's obtained from our helium-air experiments. Figure 9a gives $P(u;x)$ from the porous tube experiments. The nearly symmetric character of the distribution is noted. This is to be contrasted with the distributions of concentration shown in Fig.9b; these display dramatically the effect of boundedness.

The need for compromise in selection of bin size is exaggerated when joint probability density functions are developed. Suppose we consider two time series, e.g., in the u-velocity and c. We can form the joint probability density function $P(u,c;\mathbf{x})$ by dividing the area encompassing the ranges of the two variables into rectangles. The two time series are then scanned simultaneously and pairs of u and c deposited in the appropriate two-dimensional bins. Again the population in each bin is made a percent. Smoothing this case involves the development of contoured surfaces corresponding to constant values of $P(u,c;\mathbf{x})$.

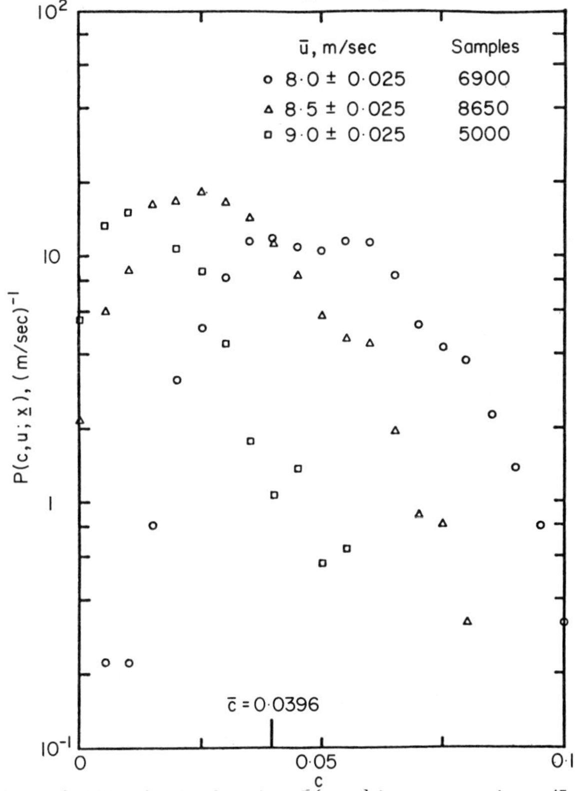

Fig.10 The joint probability density function $P(c,u;\mathbf{x})$ in a porous pipe: $x/L = 0.75$, $r/R = 0.75$.

We show in Fig.10 an example of a joint probability of u and c taken from [30] in the form of slices of the three-dimensional surface $P(c,u;\mathbf{x})$ parallel to the $P\!-\!c$ axes. The resultant contours show that the fluid with the higher velocity has a lower concentration of helium; in fact $\overline{u'c'}/(\overline{u'^2}\,\overline{c'^2})^{1/2} = -0.78$ for all data used in constructing this distribution.

Predictive methods for turbulent flows based on direct analysis of the probability density functions, single, joint and conditioned, are not well developed at present. Although such approaches are appealing in that they deal with fundamental quantities, there are a variety of assumptions required to carry through the analysis for even relatively simple flows. The situation has been carefully reviewed by Dopazo and O'Brien [31]. There is no question about the need for experimental data on the pdf's of the sort we describe here in order to provide essential guidance to theoretical developments. O'Brien* has particularly called for data on the conditioned probability $P(u_2\,|\,c;\mathbf{x})$ and its integral $\int u_2 P(u_2\,|\,c;\mathbf{x})\,du_2$.

d. Conditioning on the concentration. One of the most important recent developments in experimental turbulence research is that termed "conditioning" (cf. Kovasznay *et al.* [32]). Its development was a result of the realization (at least by the experimentalist in the laboratory if not by the theoretician developing predictive methods) that many turbulent shear flows involve a structure which is obscured by the usual averaging techniques. The most obvious case calling for special treatment relates to the existence of separate but contiguous irrotational and turbulent regions and the associated phenomenon of intermittency discussed briefly in connection with Fig.1. A hot-wire located at a point in the flow where the flow is sometimes turbulent and sometimes irrotational, yields a signal with a definite structure which is masked when the conventional averaging is performed. However, if by some discriminating technique, the periods when the signal from the hot-wire arises from the irrotational flow and the remaining periods when it is associated with turbulent flow are established, it is possible to determine, for example, separate statistics for the two "conditions", the percentage of time the flow is turbulent, and the mean values of the flow variables at the fronts when the interface crosses the hot-wire. The conditioning associated with discrimination makes available a wide variety of information related to the detailed physics of the turbulence. Discrimination has been applied to a variety of different flows, of flow variables and for a variety of purposes.

Data from our helium-air experiments can be constructively analyzed from this point of view. As we shall show in Sec.3 it is possible, under certain idealized conditions, to consider an oscillating surface of constant concentration, $c = c_o$, to correspond to a reacting surface. In developing a theory for such reacting flows, it might be useful to have experimental data on the statistical geometry of such surfaces and on the fluid mechanics at the interface. It is readily shown that, by conditioning on a particular level of helium concentration, our helium-air data can be analyzed to provide such information.

The procedure involved is as follows. The time series in $u(t;\mathbf{x})$ and $c(t;\mathbf{x})$ for a given spatial point $\mathbf{x}$ is scanned; a discriminating function $I(t;\mathbf{x})$ is established

---

* Private communication.

such that $I = 1$ when $c \geqslant c_o$; $I = 0$ when $c < c_o$. With $I(t;\mathrm{x})$ so determined, "zone" averages can be computed; for example,

$$\bar{u}_{c \geqslant c_o} = \lim_{T \to \infty} \frac{1}{T} \int_0^T u(t;\mathrm{x})\, I(t;\mathrm{x})\, dt$$

$$\bar{u}_{c < c_o} = \lim_{T \to \infty} \frac{1}{T} \int_0^T u(t;\mathrm{x})\, (1 - I(t;\mathrm{x}))\, dt$$

$$u_{c \geqslant c_o} = \lim_{T \to \infty} \frac{1}{T} \int_0^T c(t;\mathrm{x})\, I(t;\mathrm{x})\, I(t;\mathrm{x})\, dt$$

etc.. In addition $I(t;\mathrm{x})$ can be used to identify the times associated with "crossings", i.e., when $I(t;\mathrm{x})$ changes value. In fact, two sequences of times can be established, one corresponding to changes of I from 0 to 1 and the second to changes from 1 to 0. These crossings are sometimes called "fronts" and "backs" although "upstream" and "downstream" are usually less ambiguous labels.

To give an impression of the results obtained from this technique, Fig.11 shows the results for $x/L = 0.75$ obtained by conditioning on a contour with $c = c_o = 0.001$ which can be considered the air~dilute-helium interface in the porous-tube experiment. We see the distributions of the percentage of time that $c > c_o$, denoted $\bar{I}$, and of the crossing frequency f_I. We discuss first the $\bar{I}$ results. We see that near the porous wall the mixture nearly always has a concentration greater than c_o but that near the axis of the pipe it rarely does so. We discuss later the curve fitting of these experimental data.

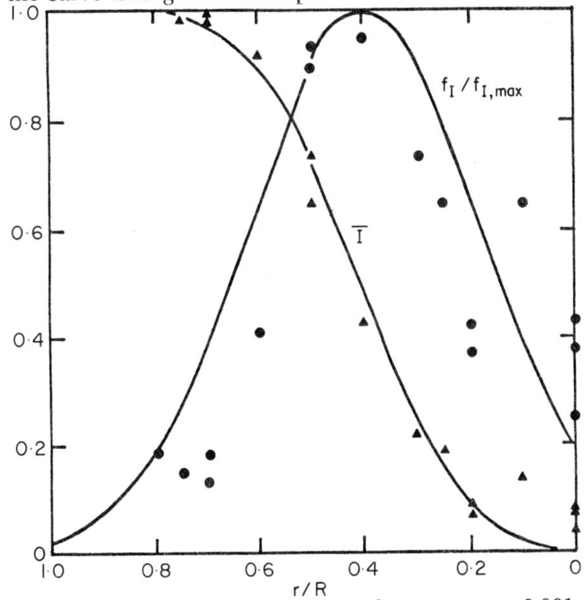

Fig.11 The distribution of intermittency and crossing frequency: $c_o = 0.001$, $x/L = 0.75$.

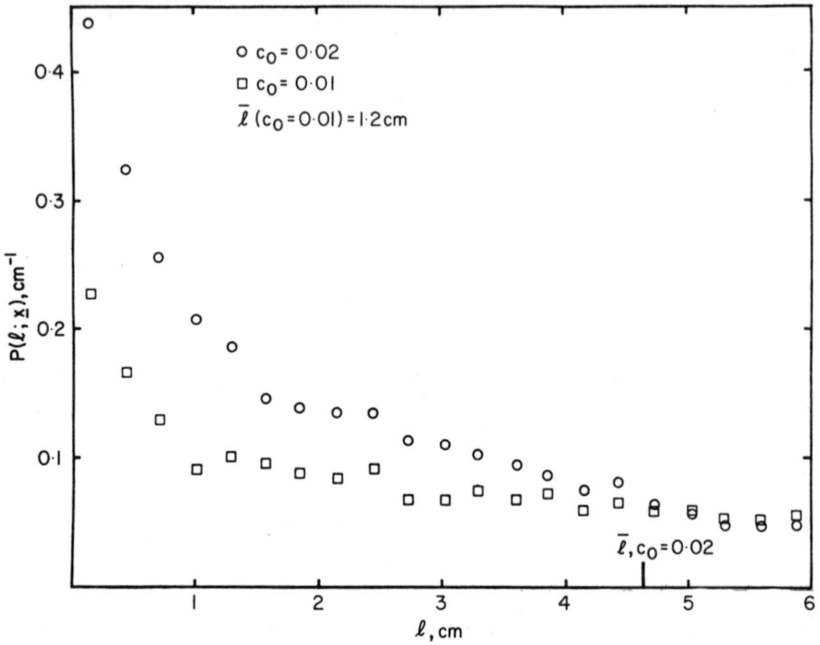

Fig.12 The probability density function for the time lengths for $c > c_o = 0.01$; $x/L = 0.75$, $r/R = 0.75$.

Information on the statistical geometry of the contour $c = c_o$ can also be obtained from the discriminating function $I(t;x)$. For example, one might want information about the lengths of time during which c is greater than or equal to c_o; for example, the way the time intervals when $c \geqslant c_o$ are distributed about the mean interval or what the crossing frequency of the contour might be. Some results relating to these features appear in Figs 11 and 12. In Fig.11 the crossing frequency f_I normalized by an estimate of its maximum value is given; a typical bell-shaped distribution is seen, as would be expected on physical grounds. The solid lines in Fig.11 derive from a theory for the frequency distribution based on the classical work of Rice [33] on the statistics of level crossings. In the present context Rice's theory concerns the distributions of intermittency, $\bar{I}(r/R; x/L)$, and of crossing frequency, $f(r/R; x/L)$, if the positions of the contour $c = c_o$ follow a normal distribution. It is not possible to check directly the validity of this assumption for our experiment but the consequences can be compared with the intermittency and crossing frequency. Certain parameters have been selected in the theory to fit our measured $\bar{I}$-distribution; the distribution of the crossing frequency is then predicted. Given the nature of the experiment and theory, the agreement is satisfactory. We show later how the crossing frequency enters the consideration of turbulent reacting flows.

Information about the intermittency and crossing frequency does not provide a complete picture of the statistical behaviour of a particular isoconcentration contour. Another element is the probability density function of the lengths of time for which $c > c_o$ and $c < c_o$ at a given spatial point. In a perfectly regular geometric pattern, only one time interval would be observed; since this is not the nature of turbulent flows, it is of interest to investigate how these time intervals are distributed. By invoking Taylor's hypothesis of a "frozen pattern", we can interpret these "time lengths" in terms of space lengths. This conversion has been made in Fig.12 where we show typical pdf's for two levels, $c_o = 0.01, 0.02$. For the lower gate level and beyond a length of one centimeter, all lengths over a wide band are nearly equally probable. For the higher values of c_o the more common behaviour corresponding to a monotonic decrease in lengths is seen. We note that because of digitization we can resolve neither the peak in $P(\ell;x)$ nor its expected decline to zero as the lengths go to zero.

e. Miscellaneous results and concluding remarks

The reader will have gathered from this discussion of the various results obtainable from our helium-air research that the measurement of the transverse velocity component (in addition to the streamwise component) would greatly expand the information gained from an experiment. The price paid is the increased complexity associated with extending the calibration procedure to include the expected range of flow deflections and a more complex data reduction. Thus the usual X-wire calibration has to be carried out in each of the several mixtures of known helium concentration. In reducing the data we use the procedures described above to determine the time series in u_1 and c; for a given u_1-c pair we use the voltage from the normal film and the voltage from the third sensor, a swept film, to determine the associated flow deflection φ. The results of experiments utilizing this technique are, *inter alia*, experimental determination of the transverse fluxes of momentum and concentration, quantities which are important in turbulence theory.

One of the disappointments of our effort described here is that the technique does not seem to have been picked up and used by others. Perhaps the emphasis we have placed on attention to the details of calibration, data collection and reduction has deterred potential users. It is also possible that the amount of electronic instrumentation required has appeared forbidding. However, now that the technique has been demonstrated and validated, only a portion of the checks we have used need be routinely applied. In addition the electronic equipment is typical of that required by most modern experimental turbulence research; the only additional equipment needed relates to the high pressure bottles necessary for the calibration gases. Thus we hope that the technique of extended hot-wire anemometry described in this section will be taken up by others; for it does seem to provide a means for obtaining valuable data on variable density effects. Our own future efforts will include work to exploit the technique in a variety of flows and thereby to build up a catalogue of data on helium-air mixing.

An interesting extension of the present work would be the study of the mixing of two gases whose molecular weights differ by more than those of helium and air. The porous tube provides a relatively simple flow situation for such an investigation. Helium could be introduced through the open end of the porous tube and one of the carbon fluoride compounds injected through the porous cylindrical surface. In this way it would appear feasible to obtain values of the parameter w (which appeared in equations (9) and (10) as high as 30. This would appear to open an entirely new domain of variable-density turbulence worthy of investigation.

3. TURBULENT REACTING FLOWS INVOLVING FAST CHEMISTRY

In providing a background to turbulent reacting flows Sec.1 introduced the idealization of fast chemistry. Here we develop the physical and mathematical ideas involved. Exposition is facilitated by considering first the general case where significant heat release leads to variable-density effects; next the case of highly dilute reactants is examined, and finally the extension of that approach to the heat-release case is considered.

3.1 The Describing Equations

We consider a turbulent flow which is two-dimensional in the mean, employing the usual subscript notation for the coordinates x_i and velocity components u_i. Favre averaging is used to secure, as discussed in Sec.2, the simplest appearance of the describing equations. Finally, attention is limited to flows in which a boundary-layer approximation is valid, i.e., to flows where the derivatives with respect to x_2 are much greater than those with respect to x_1, and where the mean velocity $\bar{u}_1$ is much greater than $\bar{u}_2$. Careful development of these equations has been provided by Bray [34].

We make several additional assumptions which simplify the analysis. The flow is assumed to be low-speed so that the static enthalpy represents essentially all of the energy in the flow and so that the static pressure fluctuations are thermodynamically unimportant.

In addition to the equations which must be introduced to effect closure of the resulting equations, others are needed to complete the description of the thermodynamic behaviour; these are a static enthalpy-temperature-composition relation, an equation of state, and the equations relating the species and element mass fractions. These may be written:

$$\tilde{h} = c_p \sum_{i=1}^{N} \left(\tilde{T}\frac{\tilde{Y}_i}{W_i} + \tilde{Y}_i \Delta_i\right),$$

$$\bar{p} = \bar{\rho} R_o \tilde{T} \sum_{i=1}^{N} \frac{\tilde{Y}_i}{W_i},$$

$$\tilde{Z}_i = \sum_{j=1}^{N} r_{ij}' \tilde{Y}_j \qquad i = 1, M,$$

where Z_i is the mass fraction of element i; c_p is the coefficient of molar specific heat at constant pressure, taken to be a constant; Δ_i is the constant in the static-enthalpy-temperature relation for species i, representing the heat of formation; R_o is the universal gas constant and r_{ij} is the number of atoms of element i in a molecule of species j times the ratio of the molecular weights W_i/W_j. Note that these remarks on thermodynamic preliminaries already expose difficulties due to turbulence if conventional averaging is used.

The conservation equations themselves may be written as follows:

continuity:

$$\frac{\partial}{\partial x_i}(\bar{\rho}\,\tilde{u}_i) = 0 \tag{14}$$

x_1-momentum:

$$\frac{\partial}{\partial x_i}(\bar{\rho}\,\tilde{u}_i\tilde{u}_1) = -\frac{\partial\bar{p}}{\partial x_1} - \frac{\partial}{\partial x_2}(\overline{\rho u_1'' u_2''}) \tag{15}$$

static enthalpy:

$$\frac{\partial}{\partial x_i}(\bar{\rho}\tilde{u}_i\tilde{h}) = -\frac{\partial}{\partial x_2}(\overline{\rho u_2'' h''}) \tag{16}$$

species:

$$\frac{\partial}{\partial x_i}(\bar{\rho}\tilde{u}_i\tilde{Y}_j) = -\frac{\partial}{\partial x_2}(\overline{\rho u_2'' Y_j''}) + \bar{\dot{w}}_j, \quad j = 1, \ldots N \tag{17}$$

element mass fraction:

$$\frac{\partial}{\partial x_i}(\bar{\rho}\tilde{u}_i\tilde{Z}_j) = -\frac{\partial}{\partial x_2}(\overline{\rho u_2'' Z_j''}), \qquad j = 1, \ldots M \tag{18}$$

turbulent kinetic energy:

$$\frac{1}{2}\frac{\partial}{\partial x_i}(\bar{\rho}\tilde{u}_i\tilde{q}) + \overline{\rho u_1'' u_2''}\frac{\partial u_1}{\partial x_2} = -\left(\frac{1}{2}\frac{\partial}{\partial x_2}(\overline{\rho u_2'' u_i'' u_i''}) + \overline{u_i''\frac{\partial p}{\partial x_i}}\right) - \Phi \tag{19}$$

species intensity:

$$\frac{1}{2}\frac{\partial}{\partial x_i}(\bar{\rho}\tilde{u}_i\tilde{I}_j) + \overline{\rho u_2'' Y_j''}\frac{\partial\tilde{Y}_j}{\partial x_2} = -\frac{1}{2}\frac{\partial}{\partial x_2}(\overline{\rho u_2'' Y_2''})^2 - X_j \tag{20}$$

Other equations, for example, that for the Reynolds stress $\overline{\rho u_1'' u_2''}$ follow a similar pattern. Since most of the notation is standard, only a few of the quantities in the above equations need be discussed in detail. The Favre-averaged turbulent kinetic energy is $\tilde{q} = \overline{\rho u_i'' u_i''}/\bar{\rho}$; the intensity of the concentration fluctuations is denoted as $\tilde{I}_j = \overline{\rho Y_j'' Y_j''}/\bar{\rho}$. These two quantities indicate the degree of heterogeneity of the velocity and concentration fields respectively. The dissipation of these two quantities due to molecular effects is contained in Φ and X_i, the so-called dissipation and scalar dissipation respectively.

To close this set of equations in a predictive method requires further analysis at two levels. Strictly fluid mechanical information is required to eliminate the flux terms, $\overline{\rho u_1'' u_2''}$, $\overline{\rho u_2'' h''}$, $\overline{\rho u_2'' Y_j''}$, $\overline{\rho u_2'' Z_j''}$, $\overline{\rho u_2'' u_i'' u_i''}$, and $\overline{\rho u_2'' Y_j''^2}$ and the pressure velocity correlations as contained in $\overline{u_i'' \partial p / \partial x_i}$; this information is the same as that required in the treatment of the dispersion of passive species, i.e., of non-reacting flows. As we have indicated earlier, this does not imply that that way to eliminate these extra terms, especially for flows with significant density variations, is fully understood and generally accepted. Nevertheless, for the flux terms, straightforward application of the modelling practices, reasonably validated for turbulent flows with constant properties, has the virtue of simplicity, hopefully without excessive inaccuracy. In this regard the writer [35] has shown that for low speed flows a relatively simple closure scheme based on the Prandtl-Kolmogorov notion of a constant turbulent Reynolds number* predicts well the insensitivity of the spreading angle in a two-dimensional mixing region to the ratio of the densities in the two streams produced by differences in composition. For the case of high Mach numbers this was not found to be true. We are unaware of any other studies attempting to assess the modelling approaches for Favre averaging or in fact for conventional averaging when significant density fluctuations must be taken into account.

The other level of closure concerns chemical effects and is confined to the terms involving the mean rate of creation, i.e., $\overline{w}_i$. If these terms could be convincingly modelled, the phenomenology of turbulent reacting flows would correspond to that for flows with several passive scalars, i.e., to turbulent flows of great complexity but with no more fundamental difficulties than are already treated in some fashion. The creation terms add further difficulties.

In this situation it has been customary for engineers to replace $\overline{w}_i$ by the appropriate products of mean values; for example, if $w_1 = -k\rho^2 Y_1 Y_2$, we have rigorously

$$\overline{w}_1 = -\overline{\rho^2}\,\tilde{k}\,\tilde{Y}_1\,\tilde{Y}_2 \left(1 + \frac{\overline{\rho^2 Y_1''}}{\overline{\rho^2}\,\tilde{Y}_1} + \frac{\overline{\rho^2 Y_2''}}{\overline{\rho^2}\,\tilde{Y}_2} + \frac{\overline{\rho^2 k''}}{\overline{\rho^2}\,\tilde{k}} + \ldots\right) \tag{21}$$

plus additional terms involving higher order correlations. The crude approximation, which neglects the effect of turbulence on the mean rate of creation, is

$$\overline{w}_i = -\overline{\rho^2}\,\tilde{k}\,\tilde{Y}_1\,\tilde{Y}_2 \tag{22}$$

i.e., the multiplier in equation (21) is replaced by unity. A major effort in turbulence research on turbulent reacting flows is directed at incorporating at least some of the turbulence effects suggested by equation (21) into the reaction modelling.

Because of the complexities involved only relatively simple chemical systems have so far been considered in attempting to include turbulence effects. The standard case for nonpremixed reactants is described by the chemical equation given by equation (4), i.e.,

---

* The turbulent Reynolds number is defined as $\tilde{q}^{1/2}\ell/\nu_T$ where ℓ is a length scale of the energy-containing eddies and ν_T is the turbulent eddy viscosity.

$$M_1 + M_2 \rightarrow M_3$$

Thus, two reactants undergo a one-step unidirectional reaction to form one product; thus $N = 3$ in equation (17). Furthermore, it is usually assumed that two elements are involved in the reactants but that an inert diluent, which may be treated as a third element, exists; thus $M = 3$ in equation (18). This idealized system can be considered to represent a simple fuel-oxidizer pair reacting to form a single product.

3.2 Dilute Reactants

A method for analysing the above chemical system in the case of "fast chemistry" in terms of the probability density functions for a passive scalar is now described. Initially we consider the case of highly dilute reactants in an isothermal, uniform background gas; the fluid mechanical and chemical aspects are then uncoupled in the sense that the heat release does not affect the turbulence but the turbulence does, of course, affect the effective chemical rates. We shall subsequently discuss more general cases in the context of pdf's.

Let us consider a two-dimensional mixing layer such as shown in Fig. 1a; the pressure is taken to be constant. The concentration of M_1 in the faster moving stream is denoted by Y_{11}, that of M_2 in the slower moving stream by Y_{22}. A natural parameter relating these two concentrations is $\hat{Y} \equiv W_1 Y_{22}/W_2 Y_{11}$ where W_i is the molecular weight of species i. Similar convenient parameters arise in other flows.

We introduce a new variable, which is a linear combination of the two element mass fractions and thus a conserved scalar and which we denote $\zeta(\mathrm{x};t)$. We have

$$\zeta \equiv (Z_1/Y_{11}) - \hat{Y}(Z_2/Y_{22}) \tag{23}$$

so that $\zeta = 1$ in the faster moving stream, and $-\hat{Y}$ in the slower. From the relations between the species and element mass fractions equation (23) also yields

$$\zeta = (Y_1/Y_{11}) - \hat{Y}(Y_2/Y_{22}) \tag{24}$$

so that $\zeta = 0$ when $Y_1 = Y_2 = 0$.

If now we invoke the "fast chemistry" assumption, namely that $Y_1(\mathrm{x},t)\, Y_2(\mathrm{x},t) = 0$, we see that $\zeta(\mathrm{x},t)$ determines both $Y_1(\mathrm{x},t)$ and $Y_2(\mathrm{x},t)$ as follows: if $\zeta(\mathrm{x},t) > 0$, $\zeta = Y_1/Y_{11}$; if $\zeta(\mathrm{x},t) < 0$, $\zeta = -\hat{Y}(Y_2/Y_{22})$. Thus if we know the pdf of ζ, i.e., $P(\zeta;\mathrm{x})$, then

$$\begin{aligned} \overline{Y}_1/Y_{11} &= \int_0^1 \zeta P(\zeta;\mathrm{x})\, d\zeta \\ -\hat{Y}(\overline{Y}_2/Y_{22}) &= \int_{-Y}^0 \zeta P(\zeta;\mathrm{x})\, d\zeta \end{aligned} \tag{25}$$

Furthermore, with certain reasonable approximations

$$\frac{W_1 \overline{Y}_3}{W_3 Y_{11}} = \frac{1}{1+\hat{Y}} \left(\int_{-Y}^{0} (\hat{Y} + \zeta)\, P(\zeta;\mathbf{x})\, d\zeta + \hat{Y} \int_{0}^{1} (1 - \zeta)\, P(\zeta;\mathbf{x})\, d\zeta \right) \tag{26}$$

Thus we see that if $P(\zeta;\mathbf{x})$ is known, the entire composition is readily given.

The value $\zeta(\mathbf{x};t) = 0$ corresponds to the flame sheet, idealized in this treatment as a surface of discontinuity. The actual thickness of the surface depends on the instantaneous strain arising from the turbulent motions and on the chemical kinetic rate at the molecular level.* At this surface reactants are destroyed and product created. The idealization of an infinitely thin sheet leads to the view that the destruction term $\dot{w}_i(\mathbf{x},t)$ for a reactant, M_1, say, can be considered a pulse train of delta functions each with a certain strength. From this standpoint

$$\overline{\dot{w}}_1(\mathbf{x}) = -W(\mathbf{x})\, f(\mathbf{x}) \tag{27}$$

where $f(\mathbf{x})$ is the mean frequency of the pulses and $W(\mathbf{x})$ is the mean rate of destruction of M_1 at each crossing. Thus we see how the crossing frequency discussed earlier might enter into a direct modelling of the creation term. Presumably W will depend on the mean concentrations of the two reactants and on the mean turbulent strain.

Dopazo and O'Brien [31] have recently reviewed the analytic methods under development for the direct calculation of pdf's such as $P(\zeta;\mathbf{x})$. However, these methods are fraught with difficulties and require crucial assumptions that, at present, have not been verified experimentally.

Lin and O'Brien [20] show how experimental data on the pdf of passive scalars such as temperature or concentration in a given flow can be used to compute the composition in that same flow if reactants and product involved in chemical reaction were present, highly diluted of course. This seems a useful observation and one that may lead to important information about turbulent reacting flows, albeit highly idealized ones. For example, sampling that is conditioned on a specified concentration level corresponds to the flame sheet for a given value of either $\hat{Y}$ or its equivalent in other analyses.

The following approach can, in principle, be successively improved to allow an increasing amount of fluid mechanical information on the behaviour of ζ, to be incorporated. It is based on the relation between the pdf of ζ and its statistical behaviour, which may be computed from a hierarchy of conservation equations related to ζ. The starting point is the following set of equations based on the properties of $P(\zeta;\mathbf{x})$:

$$\begin{aligned} 1 &= \int_{-Y}^{1} P(\zeta;\mathbf{x})\, d\zeta \\ \overline{\zeta}(\mathbf{x}) &= \int_{-Y}^{1} \zeta P(\zeta;\mathbf{x})\, d\zeta \end{aligned} \tag{28}$$

* Williams [36] provides a current entry into the considerable literature related to the structure of the flame sheet.

$$\overline{\zeta'^2}(x) = \int_{-\hat{Y}}^{1} \zeta^2 P(\zeta;x)\,d\zeta - \bar{\zeta}^2 \qquad (28)$$

$$\overline{\zeta'^3}(x) = \int_{-\hat{Y}}^{1} \zeta^3 P(\zeta;x)\,d\zeta - 3\overline{\zeta'^2}\,\bar{\zeta} - 3\bar{\zeta}^3 \qquad \text{(cont.)}$$

These equations can be continued indefinitely. Suppose we know from the conservation equations for $\bar{\zeta}$, $\overline{\zeta'^2}$, $\overline{\zeta'^3}$, etc. the values of these quantities at a particular space point and that we assume a form for $P(\zeta;x)$, containing a number of unassigned coefficients. For example, we might adopt the following parametric form capable of representing the correct general characteristics of pdf's:

$$P(\zeta;x) = \delta_+(x)\,\delta(\zeta - 1) + \delta_-(x)(\zeta + \hat{Y}) + a_o e^{-a^2(\zeta-\zeta_o)^2}$$

$$(1 + a_1(\zeta - \zeta_o) + \ldots)^2 \qquad (29)$$

The $a_i = a_i(x)$, $i = 1, 2, \ldots$ and $\zeta_o = \zeta_o(x)$ are to be determined from an appropriate number of equations of the form of equation (28), and the $\delta_+(x)$ and $\delta_-(x)$ are the known strengths of delta functions at $\zeta = 1$, $\hat{Y}$ to take into account intermittency (i.e., the existence of two interfaces between the external, irrotational fluid and the turbulent fluid).

The assumed form has several features that should be noted. First, if a_1 and all subsequent coefficients are zero, it corresponds to part of a Gaussian distribution with a peak value not located at $\bar{\zeta}$ and with truncation outside of the two bounds, $\zeta = 1$, $-\hat{Y}$. If ζ_o is outside these bounds, the distribution provided by equation (29) is monotonic within the range of non-zero ζ. In contrast if $\zeta_o = \bar{\zeta}$, and a^2 is suitably large, we obtain a Gaussian distribution and the bounds are inoperative. Since a_o must be positive, the quadratic squaring of the multiplying series assures that $P(\zeta;x)$ is positive definite. In Fig.13 we sketch several distributions which can be represented by equation (29) with various values for the parameters involved.

If equation (29) is substituted into equations (28) (continued as necessary) there result non-linear transcendental equations which must be solved numerically for a_o, a, ζ_o and an appropriate number of the a_i coefficients. So far the numerical examples which have been worked out involve only a_o, a, and ζ_o and have thus satisfied the first three of equations (28). We show in Fig.14 some typical results for the composition distribution across a two-dimensional mixing layer. Of particular interest is the reaction zone which is characterized by non-zero values of the means of the two reactants, i.e., $\bar{Y}_1$ and $\bar{Y}_2$. It is across this zone that the flame sheet oscillates, destroying reactants and producing product, which is subsequently diffused across the entire mixing layer.

It is informative to consider the various predicted pdf's of ζ across the mixing zone as shown in Fig.15. Only the continuous part of the pdf's are given; delta functions of appropriate strength are located at $\zeta = 1$, $-\hat{Y}$ such that the integral of the pdf over all ζ is unity. In the central region of the flow the distributions of ζ are nearly Gaussian with the bounds on ζ inoperative. This fact has been

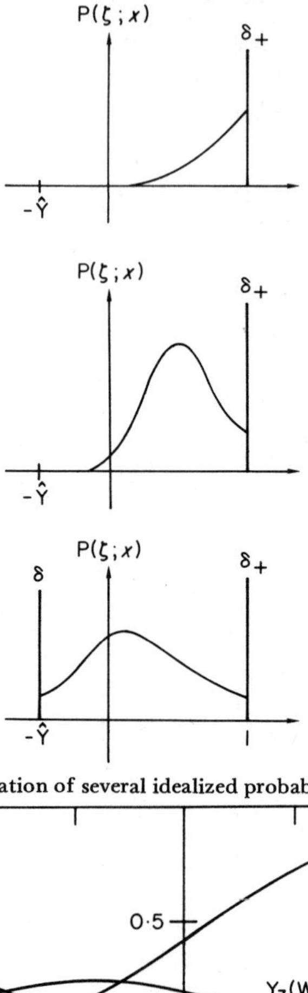

Fig.13 **Schematic representation of several idealized probability density functions, $P(\zeta;x)$.**

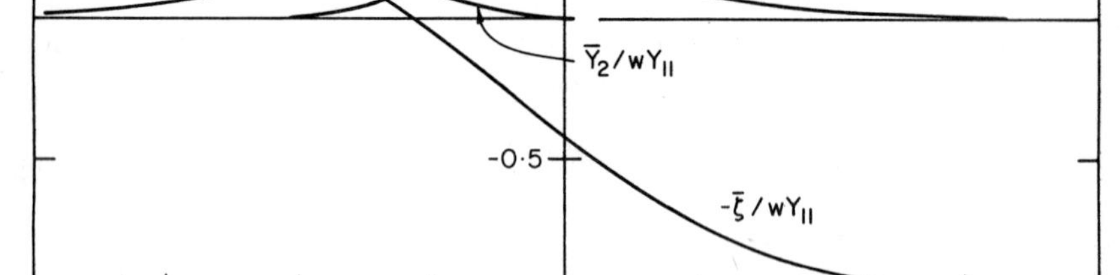

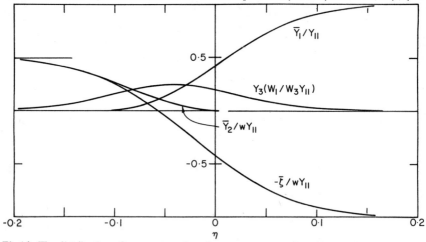

Fig.14 **The distribution of reactants and products across a two-dimensional mixing layer, $\hat{Y} = 0.5$.**

used by Alber and Batt [24] in their analysis of reaction in a mixing layer. Outside of this central core the limiting bounds on ζ truncate the distributions and result in the asymmetries shown.

It is important to note that with this approach predictions (or estimates) are

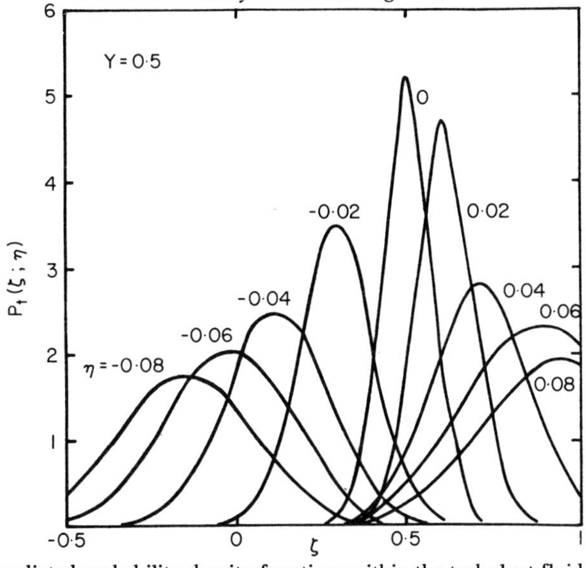

Fig.15 **The predicted probability density functions within the turbulent fluid across a two-dimensional mixing layer: $\hat{Y}$ = 0.5.**

needed of the intermittency distribution, $\bar{I}$ (which determines δ_+, δ_-), of $\overline{\zeta'^2}$ and of as many additional fluid mechanical quantities relative to ζ as are necessary to define the coefficients in equation (29). Libby [23] has pointed out that the available models for determining the intensity of passive scalars (in this connection $\overline{\zeta'^2}$), are deficient and require further development. Our ability to predict higher moments of the fluctuations of ζ can be expected to diminish further as the degree of the moments increases. At some point, therefore, the process of successively including additional terms in the pdf and additional fluid mechanical information brings negligible additional accuracy.

The problem of solving the transcendental equations for the several coefficients in the pdf, given fluid mechanical properties, may be carried out by various numerical means. We have used "descent" methods which involve introduction of a positive definite measure of error and a systematic reduction of that error. For the case where the intermittency, the mean value of ζ and its intensity are all specified, convergence is found to be reasonably rapid; when the third moment, $\overline{\zeta'^3}$, is also specified, a new technique must apparently be used to achieve convergence; studies of this question are underway at present.

3.3 The Heat-Release Case

The ideas developed in Sec.3.2 can be extended and applied to the more interesting case of flows with significant heat release and therefore with variable density. This extension is outlined with reference to the same two-dimensional mixing layer considered earlier only now within the more general framework of finite chemical kinetics in order to show the role of joint pdf's introduced earlier. The same parameters define the reactant concentrations in the two streams and the resultant concentration parameter, $\hat{Y}$. Moreover, the definitions of ζ in terms of element and species mass fractions in equations (24) and (25) still pertain. However, we must introduce as additional parameters the static enthalpies of the two streams, h_1 and h_2.

With some plausible assumptions, the state of the mixture at a given space-time point can be shown to be determined by the instantaneous values of ζ and one species concentration.* For present purposes we take the latter to be Y_1. The following relations prevail:

$$
\begin{aligned}
\frac{h}{h_1} &= \frac{1}{1+\hat{Y}}\left(\hat{Y} + \frac{h_2}{h_1} + \left(1 - \frac{h_2}{h_1}\right)\zeta\right) \\
Y_2 &= w(Y_1 - Y_{11}\zeta) \\
Y_3 &= \frac{W_3}{W_1}\left(\frac{Y_{11}}{1+\hat{Y}}(\hat{Y}+\zeta) - Y_1\right) \\
\frac{\rho}{\rho_1} &= \left(\frac{h}{h_1} - Y_3\left(\frac{\Delta}{h_1}\right)\right)^{-1} \\
\frac{W_1}{W} &= \frac{W_1}{W_4} + \frac{Y_{11}\hat{Y}}{1+\hat{Y}}\left(1 - \frac{W_3}{W_4}\right) + Y_1 - \zeta Y_{11}\left(1 - \frac{W_2}{W_4} - \frac{1}{1+\hat{Y}}\left(1 - \frac{W_3}{W_4}\right)\right) \\
\frac{T}{T_1} &= \frac{1}{\rho/\rho_1}\,\frac{1}{W_1/W}
\end{aligned}
\qquad (30)
$$

In these equations the heat of formation of the inert diluent and of the reactants has been assumed to be zero and that of the product to be Δ.

An examination of equations (30) exposes some new and interesting situations arising in flows with significant heat release. We note that in addition to the single parameter, $\hat{Y}$, involved in the case of negligible heat release, the thermodynamic state depends on various molecular weight ratios; on the enthalpy ratio, h_2/h_1; and on a heat release parameter, Δ/h_1. The molecular-weight ratios are determined by the nature of the gases involved but the latter two parameters are determined by the flow situation and suggests a variety of possibilities. For example, if we modify the last parameter somewhat and assume $(\Delta Y_{3,\max}/h_1) \ll 1$ where $Y_{3,\max}$ denotes the maximum value of Y_3, heat release will be unimportant even though the density might still be highly variable because of enthalpy heterogeneities. This situation arises in the case where a chemical reaction of trace species (pollutants for example) which is not energetically significant, occurs in a non-uniform turbulent background gas.

* Williams [37] and Bilger [38] have been instrumental in demonstrating this fact which results from a systematic application of the so-called Shvab-Zeldovitch coupling functions.

Another limiting case of interest arises if $(h_2/h_1) = 1$ but $(\Delta Y_{3,max}/h_1) \gg 1$. In this case the variability of the density is due entirely to heat release and would be entirely absent were it not for chemical reaction. The case of uniform density described earlier corresponds to $(h_2/h_1) = 1$ and $(\Delta Y_{3,max}/h_1) \ll 1$.

Consistent with the chemical system given by equation (23), we have in general for the instantaneous destruction of M_1,

$$\dot{w}_1 = -\rho^2 A e^{-T_a/T} Y_1 Y_2 \tag{31}$$

where A is a constant which in SI units has dimensions $(m^3/kg/s)$ and T_a is an activation temperature. The above results imply that the mean of $\dot{w}_1$ may be expressed as

$$\bar{\dot{w}}_1(\mathrm{x}) = -\rho_1^2 A w \int_{-\hat{Y}}^{1} \int_0^{Y_{11}} (\rho/\rho_1)^2 e^{-(T_a/T_1)(T_1/T)} Y_1(Y_1 - Y_{11}\zeta) P(Y_1,\zeta;\mathrm{x})\, dY_1\, d\zeta \tag{32}$$

where $P(Y_1,\zeta;\mathrm{x})$ is the joint probability density of Y_1 and ζ at $\mathbf{x}$. We thus see how the joint pdf's discussed earlier enter in the problem of chemical reaction; here we need the pdf of two scalars, one passive, one active. There appear to be no data on such functions, although the simultaneous measurement of helium concentration and of temperature would be feasible and might provide information of value in developing suitable models for $P(Y_1,\zeta;\mathrm{x})$. We plan such measurements in the near future.

Equation (32) serves to emphasize the complexities of modelling chemical reactions involving finite-rate chemistry and heat release even in a relatively simple chemical system. The assumption of "fast chemistry" leading to the non-coexistence of the two reactants reduces the joint pdf to a single one in ζ of the following form.

$$\begin{aligned} P(Y_1,\zeta;\mathrm{x}) &= P(\delta(Y_1 - Y_{11}\zeta), \zeta;\mathrm{x}), && 0 \leqslant \zeta \leqslant 1 \\ &= P(\delta(Y_1), \zeta;\mathrm{x}), && -\hat{Y} \leqslant \zeta < 0 \end{aligned} \tag{33}$$

The supposition that $Y_1(\mathrm{x},t)\, Y_2(\mathrm{x},t) = 0$ requires also that $A \to \infty$ for otherwise $\bar{\dot{w}}$ would, from equation (32), be zero. Of course $\bar{\dot{w}}$ cannot be determined from equation (32). Equations (30) apply separately to the two ranges of ζ so that if $P(\zeta;\mathrm{x})$ is known the entire mean state of the gas at $\mathbf{x}$ is readily determined.

In terms of Favre averages the generalization of equations (26) and (27) leads to

$$\begin{aligned} \widetilde{Y}_1(\mathrm{x}) &= Y_{11}(\rho_1/\bar{\rho}) \int_0^1 (\rho/\rho_1) P(\zeta;\mathrm{x})\, d\zeta \\ \widetilde{Y}_2(\mathrm{x}) &= -w Y_{11}(\rho_1/\bar{\rho}) \int_{-\hat{Y}}^0 (\rho/\rho_1) P(\zeta;\mathrm{x})\, d\zeta \end{aligned} \tag{34}$$

$$\frac{W_1}{W_3}\frac{\widetilde{Y}_3(\mathrm{x})}{Y_{11}} = \frac{\rho_1}{\rho}\frac{1}{1+Y}\left(\int_{-\hat{Y}}^{0}(\hat{Y}+\zeta)(\rho/\rho_1)\,P(\zeta;\mathrm{x})\,d\zeta + \int_{0}^{1}(1-\zeta)(\rho/\rho_1)\,P(\zeta;\mathrm{x})\,d\zeta\right) \tag{34 (cont.)}$$

where $(\bar{\rho}/\rho_1)$ is obtained from

$$\frac{\bar{\rho}}{\rho_1} = \int_{-\hat{Y}}^{0}(\rho/\rho_1)\,P(\zeta;\mathrm{x})\,d\zeta + \int_{0}^{1}(\rho/\rho_1)\,P(\zeta;\mathrm{x})\,d\zeta \tag{35}$$

The pdf of ζ can be modelled in the same fashion as indicated earlier in connection with the case of negligible heat release. However, if the benefits of Favre averaging are to be realized, the parameters in $P(\zeta;\mathrm{x})$ must be related to $\tilde{\zeta}$ and $\overline{\rho\zeta''^2}/\bar{\rho}$, etc., by the appropriately generalized forms of equations (28):

$$\begin{aligned} \tilde{\zeta} &= (\rho_1/\bar{\rho})\int_{-\hat{Y}}^{1}(\rho/\rho_1)\,\zeta\,P(\zeta;\mathrm{x})\,d\zeta \\ \overline{\rho\zeta''^2}/\bar{\rho} &= (\rho_1/\bar{\rho})\int_{-\hat{Y}}^{1}(\rho/\rho_1)\zeta^2\,P(\zeta;\mathrm{x})\,d\zeta - \bar{\rho}\tilde{\zeta}^2 \end{aligned} \tag{36}$$

for example; obvious additions to this hierarchy can be made as needed.

We thus see that the variable density, which appears throughout equations (14)–(20) and in equations (36) and their extensions, couples the coefficients in the model for $P(\zeta;\mathrm{x})$ to the mean velocity and to $\tilde{\zeta}$, $\overline{\rho\zeta''^2}$, etc.. The effort needed to exploit this approach remains for the future. Clearly some interesting effects can be anticipated.

3.4 *Concluding Remarks*

This review of the research of the writer and his colleagues on the problem of chemically reacting turbulent flows, has attempted to indicate the complexities involved and the importance of close interrelation between experimental and theoretical work. Clearly much research remains to be carried out. Even though the simplest case of negligible heat release with fast chemistry warrants further study, it seems advisable to pursue the approach to the heat release case outlined here, since it does correspond to the interesting situation wherein the dynamic coupling between chemistry and turbulence is complete; not only does the turbulence affect the effective chemical rates, but the heat release alters the turbulence as well. It will thus complement for non-premixed reactants the recent work of Bray and the author referred to earlier.

The assumption of fast chemistry does of course exclude consideration of a wide variety of important questions concerning ignition and extinction. These should probably be approached systematically starting with the treatment of negligible heat release in a heterogeneous, non-isothermal background diluent as suggested earlier, followed by consideration of cases of increasing complexity. In this context the experimental effort should provide a variety of data related to turbulent shear flows with significant variable density effects including the behaviour of scalars in such flows.

ACKNOWLEDGEMENTS

The research reported here was supported by the Office of Naval Research under Contract N00014-0226-005 (Subcontract No.4965-26) as part of Project SQUID. The author gratefully acknowledges the assistance of Dr. John LaRue and Mr. Richard Stanford.

REFERENCES

1. Bertram, M.H. (ed). "Compressible Turbulent Boundary Layers", NASA SP-216, 1969.
2. Morkovin, M.V. *In* "The Mechanics of Turbulence", pp.367-380, Gordon and Breach, New York and London, 1964.
3. Hinze, J.O. "Turbulence", pp.430-431, McGraw-Hill, New York, 1959.
4. NASA. "Free Turbulent Shear Flows", NASA SP-321, 1973.
5. Kovasznay, L.S.G. "The Hot-Wire Anemometer in Supersonic Flow", *J. Aero Sci.* **17**, 565-572, 1950.
6. Bradshaw, P. "An Introduction to Turbulence and its Measurement", pp.155-164, Pergamon Press, Oxford, 1971.
7. Gibson, C.H. "Digital Techniques in Turbulence Research", AGARDograph No.174, 1973.
8. van Atta, C.W. *In* "Annual Review of Fluid Mechanics", Vol.6, pp.75-91, Annual Reviews, Palo Alto, 1974.
9. Sandborn, V.A. "A Review of Turbulence Measurements in Compressible Flow", NASA TM X-62, 337, 1974.
10. Whitelaw, J.H. "Developments in Laser-Doppler Anemometry". *In* "Studies in Convection", Vol.1, Academic Press, London, 1975.
11. Catalano, G.D. "An Experimental Investigation of an Axisymmetric Jet in a Co-Flowing Stream", M.S. Thesis, University of Virginia, 1975.
12. Favre, A. *In* "Problems of Hydrodynamics and Continuum Mechanics", pp.231-266, Society for Industrial and Applied Mathematics, Philadelphia, 1969.
13. Rubesin, M.W. and Rose, W.C. "The Turbulent Mean Flow, Reynolds Stress and Heat-Flux Equations in Mass-Averaged Dependent Variables", NASA TNX 62248, 1973.
14. Libby, P.A. and Williams, F.A. *In* "Annual Review of Fluid Mechanics", Vol.8, Annual Reviews, Palo Alto, 1976.
15. Murthy, S.N.B. (ed). "Turbulent Mixing in Nonreactive and Reactive Flows", Plenum Press, New York and London, 1975.
16. Spalding, D.B. *In* "Turbulent Mixing in Nonreactive and Reactive Flows", pp.85-130, Plenum Press, New York and London, 1975.
17. Bray, K.N.C. *In* "Analytical and Numerical Methods for Investigation of Flow Fields with Chemical Reactions, Especially Related to Combustion", II-2, AGARD Conference Proceedings No.164, 1975.
18. Hawthrone, W.R., Weddell, D.S. and Hottel, H.C. *In* "Third Symposium on Combustion, Flame and Explosion Phenomena", pp.266-288, Combustion Institute, Pittsburgh, 1949.
19. Toor, H.L. "Mass Transfer in Dilute Turbulent and Nonturbulent Systems with Rapid Irreversible Reactions and Equal Diffusivities", *A.I.Ch.E.J.* **8**, 70, 1962.
20. Lin, C.H. and O'Brien, E.E. "Turbulent Shear Flow Mixing and Rapid Chemical Reactions: An Analogy", *J. Fluid Mech.* **64**, 195, 1974.
21. Bilger, R.W. and Kent, J.H. "Concentration Fluctuations in Turbulent Jet Diffusion Flames", *Combust. Sci. and Technol.* **9**, 25-29, 1974.
22. Libby, P.A. *In* "Analytical and Numerical Methods for Investigation of Flow Fields with Chemical Reactions, Especially Related to Combustion", II-5, pp.1-18, AGARD Conference Proceedings No.164. *Also in* "Turbulent Mixing in Nonreactive Flows" (S.N.B. Murthy, ed), pp.333-365, Plenum Press, New York and London, 1974.
23. Libby, P.A. "On Turbulent Flows with Fast Chemical Reactions. Part III: Two-Dimensional Mixing with Highly Dilute Reactants", *Combust. Sci. and Technol. (in press)*, 1976.

24. Alber, I.E. and Batt, R.G. "An Analysis of Diffusion-Limited First and Second Order Chemical Reactions in a Turbulent Shear Layer", *A.I.A.A.J. (in press)*, 1976.
25. Way, J. and Libby, P.A. "Application of Hot-Wire Anemometry and Digital Techniques to Measurements in a Turbulent Helium Jet", *A.I.A.A.J.* 9, 1567, 1971.
26. Corrsin, S. "Extended Applications of the Hot-Wire Anemometer", NACA TN 1864, 1949.
27. Stanford, R.A. and Libby, P.A. "Further Applications of Hot-Wire Anemometry to Turbulence Measurements in Helium-Air Mixtures", *Phys. Fluids* 17, 1353, 1974.
28. Way, J. and Libby, P.A. "Hot-Wire Probes for Measuring Velocity and Concentration in Helium-Air Mixtures", *A.I.A.A.J.* 8, 976, 1970.
29. Favre, A. *In* "Fifth Canadian Congress of Applied Mechanics, Proceedings" (G. Dhatt, ed), University of New Brunswick Press, Fredericton, N.B., Canada, 1975.
30. Libby, P.A. and Stanford, R.A. "Remarks on a Kinetic-Theory Approach to Turbulent Chemically Reacting Flows", *Combust. Sci. and Technol. (in press)*, 1976.
31. Dopazo, C. and O'Brien, E.E. "Statistical Treatment of Non-Isothermal Chemical Reactions in Turbulence", *Combust. Sci. and Technol. (in press)*, 1976.
32. Kovasznay, L.S.G., Kibens, V. and Blackwelder, R.F. "Large Scale Motion in the Intermittent Region of a Turbulent Boundary Layer", *J. Fluid Mech.* 41, 283-325, 1970.
33. Rice, S.O. "Mathematical Analysis of Random Noise", *Bell System Tech. J.* 24, 46-108, 1945. *Also in* "Selected Papers on Noise and Stochastic Processes" (Nelson Wax, ed), pp.133-294, Dover Publications, New York, 1954.
34. Bray, K.N.C. "Equations of Turbulent Combustion. I. Fundamental Equations of Reacting Turbulent Flow. II. Boundary Layer Approximation", University of Southampton AASU Reports 330, 331, Southampton, England, 1973.
35. Libby, P.A. *In* "Free Turbulent Shear Flows. **Vol.I**, Conference Proceedings", NASA SP-321, 1973.
36. Williams, F.A. *In* "Annual Review of Fluid Mechanics", **Vol.3**, pp.171-188, Annual Reviews, Palo Alto, 1971.
37. Williams, F.A. "Analytical and Numerical Methods for Investigation of Flow Fields with Chemical Reactions, Especially Related to Combustion", **III-1**, pp.1-25, AGARD Conference Proceedings No.164, 1974.
38. Bilger, R.W. "The Structure of Diffusion Flows", *Combust. Sci. and Technol. (in press)*, 1976.

NOMENCLATURE

$a_i(\mathbf{x})$	parameters in pdf model (cf. equation (29))
$A_i(c)$, $B_i(c)$	discretized functions of concentration representing hot-wire calibration
c	mass fraction of helium
E_i	voltage of sensor i
f	crossing frequency
h	static enthalpy
I	intermittency function
I_i	intensity of fluctuations of the concentration of species i
k	reaction rate constant
ℓ	turbulence length scale
L	length of porous pipe
p	pressure
$P(\ )$	probability density function
q	turbulent kinetic energy
r	radial coordinate
R	radius of porous pipe

r_{ij}	number of atoms of element i in a molecule of species j
t	time
T	temperature, averaging time
T_a	activation temperature
u	main velocity component
u_1, u_2	velocity components in x_1, x_2 directions respectively
v	transverse velocity component
w	molecular weight ratio
$\bar{w}_i$	average mass rate of creation of species i
$W(\mathbf{x})$	mean rate of destruction of reactant M_1 at each flame sheet crossing
W_i	molecular weight of species i
x	axial coordinate
x_1, x_2, x_3	Cartesian coordinates
X	voltage of normal film squared
Y_i	mass fraction of species i
Y	voltage of wire squared
$\hat{Y}$	reactant concentration parameter
Z_i	mass fraction of element i
Δ	heat of formation of product
φ	flow inclination, $v = u \tan\varphi$
ρ	mass density
η	similarity variable
ν_T	eddy viscosity
Φ	dissipation
χ_i	scalar dissipation of species i
ζ	characteristic scalar (cf. equation (23) and (24))
$(\bar{\ })$	conventionally averaged quantities
$(\tilde{\ })$	mass-averaged quantities

MIXING, CONCENTRATION FLUCTUATIONS, AND MARKER NEPHELOMETRY

H.A. Becker
Department of Chemical Engineering, Queen's University
Kingston, Ontario

ABSTRACT

Marker nephelometry is the technique of studying concentration fields associated with processes of mixing and dilation by marking one or more of the feed-streams to a field with particles, and detecting particle concentrations by means of an optical probe based on light scatter principles. The theory of marker nephelometry is here presented in full with many new details and greater rigour. Problems of mixing and dilation are classified as one-feed, two-feed, and so forth, depending upon the number of physicochemically distinct feeds supplied to a system. The general features of these classes are analyzed and some general indices of the extent of turbulent mixing are defined. The mathematical and physical modelling of mixing, dilation, and chemical reaction in turbulent systems, including flames, is discussed, and it is shown how results obtained by marker nephelometry are to be interpreted in this context. A review is given of experimental investigations done with marker nephelometry. Some notes on apparatus for marker nephelometry are appended.

1. INTRODUCTION

The study of processes of mixing, material transfer, and chemical reaction frequently requires the characterization of fields of chemical composition. One would ideally like to be able to determine as a function of spatial position and time the mass fraction or mole fraction of each chemical species that is present in any significant amount. The experimental techniques that are available, however, are often far from adequate for this purpose. Progress then depends very much on a fruitful combination of theoretical prediction procedures and experimental observations. If at least some features of a composition field can be reliably measured, and if theory can be brought to predict these with convincing accuracy, then perhaps the predictions of theory concerning those features that are not presently measurable can be accepted with some confidence.

The oldest experimental method is the use of sampling probes to obtain a representative sample of material for chemical analysis from a given point or volume of space. This approach is particularly suitable for steady-state or slowly time-varying systems. Sampling may be conducted batch-wise or continuously. A wide variety of methods are available for the chemical analysis, for example gas chromatography, spectrophotometry, mass spectrometry and Raman spectroscopy. Problems of attaining good spatial resolution and rapid quenching of chemical reactions are particularly severe in laminar pre-mixed flames; techniques of probing have, however, been refined to a remarkable degree, and detailed studies of chemical reaction and transport phenomena have been carried out. Fristrom and Westenberg [1] have described both the experimental and theoretical approaches and have given a critical review of the results.

Methods of sampling and composition measurement with particular emphasis on combustion systems have also been examined by Leeper [2], Tiné [3], Beér and Chigier [4], and Chedaille and Braud [5]. Most recently, Bilger [6] has reviewed the performance of sampling probes in turbulent combustion.

The alternative to sampling probes is the use of what may be called *detector probes*. With such probes no sample is removed from a field; instead, the probe produces a signal, usually optical or electrical, in response to some aspect of the composition or concentration field. Typical of the quantities detected are pH, electrical conductivity, electrical capacitance, light absorption, light-scatter, fluorescence, and radiant emission.

A sampling probe, at least in principle, facilitates analysis for all the stable species in a mixture. However, in systems of laboratory or industrial scale, temporal averaging in sampling turbulent systems is usually unavoidable, and consequently little or no direct information is obtainable about turbulent fluctuations. Only when one approaches meteorological or oceanographic scales is it possible to discriminate virtually instantaneous values. It is also a feature of sampling probes that the rate of measurement is usually lower than with detector probes, and the mapping of a field can be very time-consuming.

The outstanding limitation of detector probes is that each probe usually responds well to only one aspect of a composition field. However, a mapping of the field of the detected quantity can often be carried out very speedily, making it much easier to investigate the effects of changes in operating conditions and system design. In addition, a number of probes have been developed that are capable of following turbulent fluctuations, e.g., probes based on electrical conductivity [7,10], electrical capacitance [11,12], electro-kinetic potential [13], light absorption [14,15], radiant emission [16,17], and Raman scattering [18].

The present article deals mainly with optical detection methods based on light-scatter principles, and the ways in which results obtained with such probes can be used in the interpretation of fields of turbulent mixing and chemical reaction. The technique of light-scattering photometry is long-established in the analysis of samples of liquids or gases containing polymers or suspended particles. The application of the same idea to the *in situ* characterization of concentration fields is more recent. The initial study was that of Rosenweig, Hottel and Williams [19];

it established the feasibility of the technique and indicated many of the capabilities and limitations. It was shown that the concentration of fluctuations of smoke particles marking one of the streams entering a gas mixing field could easily be characterized as to intensity (rms value), spectrum, autocorrelation function, two-point spatial correlation function, and integral spatial scale. The experimental method was subsequently refined, more measurements were added, and the theory was more fully explored by Becker, Hottel and Williams [20,21].

A considerable number and variety of investigations have been carried out by means of this technique over the fifteen years since its introduction. The technique itself has been improved and its capabilities have been expanded, primarily through progress in electronic instrumentation – most notably in the development of lasers. On the theoretical side, the science of turbulence modelling has undergone significant advances which have given new importance to the type of experimental studies that the light-scatter technique facilitates. The technique has, of course, the advantage common to purely optical probes of causing minimal disturbance of the flow. The introduction of the laser-Doppler method has made it possible to measure velocity by optical means also. Efforts are currently being made [22,23] to combine the light-scatter and laser-Doppler methods to obtain the covariance between fluctuations in concentration and velocity, a most interesting development which introduces the direct measurement of the turbulent transport of material.

The aims of the present discussion are, then:

(i) to describe the "state of the art" of the marker-scattered-light technique.

(ii) to present some new ideas on the theory and application of the technique.

(iii) to review the experimental results that have been obtained by this technique.

(iv) to examine the general interpretation of such results and their place in physical modelling, mathematical modelling, and the theory of turbulence and mixing.

2. BASIC FEATURES OF MARKER NEPHELOMETRY

1.1 Definition of the Technique

The light-scatter techniques to be discussed here are those in which:

1. One or more of the fluid streams entering a field is marked with particles which represent as faithfully as possible the transport of material of the marked stream(s). It is usually necessary that the marker particles be in the colloidal size range; in gases a colloidal suspension of particles is called an aerosol or smoke, in liquids it is a hydrosol, and in the general case, a sol. The marking may be done by seeding or it may occur naturally; in either case, the particles should not undergo significant change in any scattering characteristics within the field.

2. The marker concentration field is studied by means of an optical probe based on light-scatter principles. The probe is generally a combination of (i) an *exciter* system consisting of a light source and optics for producing an incident

lightbeam with suitable characteristics for the desired measurements, and (ii) a *detector* system consisting of a photoelectric transducer and optics which responds to light scattered by particles in a defined region of the field illuminated by the incident beam.

It is a distinguishing characteristic of this special kind of light-scatter technique that the marker particles should be virtually unchanging in all light-scattering properties (such as size, shape, and refractive index) during their journey from the point of introduction into the field to the point of measurement.† It is evident that a suitable nomenclature is needed to convey the special nature of this technique. We might briefly describe it as *a light-scatter technique for the study of mixing and dilation by means of passive marker particles.* The essence of this description is conveyed by the name-phrase *marker-scattered-light technique.*

In the practice of chemical analysis, however, the measurement of the concentration of particles by light scattering based on comparison with suspensions of known concentration is called *nephelometry*. The marker-scattered-light technique can therefore be called simply *marker nephelometry*. The optical probes that are used can be generally called *nephelometer probes*, without regard to whether the particles detected are marker particles or any other kind. This nomenclature which is convenient is based on the established practice in the general field of interest and will be used hereinafter.

2.2 *Implementation of the Technique*

In the practice of marker nephelometry, one usually wishes to detect the point-concentration of the marker, Γ^*. A nephelometer probe for this purpose is illustrated schematically in Fig.1. Light scattered by particles in the path of the incident lightbeam is collected by a lens and focussed on a slitted diaphragm. The slit passes on to a photoelectric transducer the light scattered from a short segment of the incident beam. The slit and the incident beam together define the probe control volume V. The signal from the transducer is ideally proportional to the number N of the particles in V at any instant, and N/V is ideally the instantaneous marker concentration Γ^* at the centre of V.

Since the response of appropriate phototransducers, e.g., photomultiplier tubes and photodiodes – is essentially linear in light flux and it is easy to arrange that the system frequency response be flat to very high frequencies, two of the most basic requirements of ideal response are easily satisfied. There are, however, many other sources of error that should be recognized and which one seeks to render negligible; these, and other inherent limitations, are discussed in the following Section.

---

† The same instrumentation can, of course, also be used to study processes where certain particles are active participants in the phenomena, e.g., soot formation and burnout in flames, and the formation of macromolecules in polymerization reactions. Such measurements would not, however, constitute a practice of the type of light-scatter technique given by the above definition.

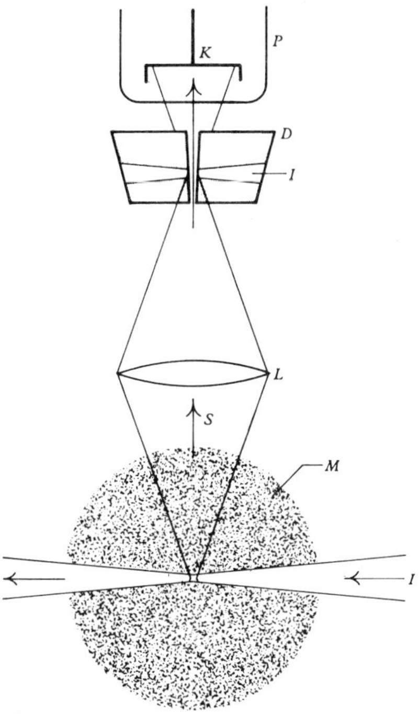

Fig. 1 The principle of marker nephelometry. *M*, mixing field; *I*, incident beam; *S*, scattered light; *D*, slitted diaphragm (field stop); *L*, lens system; *I'*, scattered-light image of incident beam on diaphragm; *P*, phototube; *K*, photocathode. (Becker, Hottel and Williams [20], their Fig.1, by permission of *J. Fluid Mech.*)

3. THEORY OF MARKER NEPHELOMETRY

3.1 *General*

The theory of marker nephelometry was first considered by Rosensweig, Hottel and Williams [19]. A more comprehensive treatment was subsequently given by Becker, Hottel and Williams [20]. Recently Shaughnessy [24] has examined factors introduced by the use of lasers. Here we summarize those points that are important in the practice of the technique. The presentation differs from the author's earlier writings, generally with a view to greater rigour and generality. Some previously unpublished material has also been added.

3.2 *Theory of Absorption and Scatter*

When radiation of spectral intensity I_λ falls upon a spherical particle through an incident solid angle $\mathbf{d\Omega}_i$, the energy flux scattered into a solid angle $\mathbf{d\Omega}_s$ in a wavelength interval $\lambda,\ \lambda + d\lambda$ is

$$\frac{1}{4\pi} A_p \eta_{\lambda s} p_\lambda I_\lambda \, d\lambda \, d\Omega_i d\Omega_s \ , \tag{3.2.1}$$

where

$A_p = \pi D_p^2/4$ = area projected by the particle in the direction of $\mathbf{d\Omega}_i$

$A_p I_\lambda d\lambda d\Omega_i$ = flux incident upon the particle in the sense of geometrical optics

$\eta_{\lambda s}$ = fraction of the incident flux that is scattered, the scattering efficiency factor

$A_p \eta_{\lambda s} p_\lambda I_\lambda \, d\lambda d\Omega_i$ = flux scattered in all directions

$p_\lambda d\Omega_s/4\pi$ = fraction of the scattered flux that is directed into $\mathbf{d\Omega}_s$.

The product $A_{\lambda s} \equiv A_p \eta_{\lambda s}$ is the *scattering cross-section* of the particle. For spherical particles, the *phase function* p_λ is obviously independent of the particle orientation, and its directional dependence involves only the angle θ between the vectors $\mathbf{d\Omega}_i$ and $\mathbf{d\Omega}_s$; in general,

$$p_\lambda = p_\lambda(\lambda, \theta, D_p, m)$$

where $m \equiv n(1 - j\kappa)$ is the complex index of refraction (generally a function of λ), n is the real refractive index, $j \equiv \sqrt{-1}$, and κ is the absorption index. The general functional form of the scattering efficiency is

$$\eta_{\lambda s} = \eta_{\lambda s}(\lambda, D_p, m) \, .$$

The scattering efficiency is not constrained to be smaller than unity; a particle may interact with radiation over a distance considerably greater than its diameter, and values of 6 or more are possible [28]. It is predicted that generally

$$\eta_{\lambda s} \to 0 \quad \text{as} \quad \pi D_p/\lambda \to 0 \, ,$$

$$\eta_{\lambda s} < 1 \quad \text{for} \quad \pi D_p/\lambda << 1,$$

$$1 < \eta_{\lambda s} < 2 \quad \text{as} \quad \pi D_p/\lambda \to \infty \, .$$

For spheres whose circumference is large compared to the wave length of light, $\pi D_p >> \lambda$, the behaviour of $\eta_{\lambda s}$ and p_λ can be effectively predicted by geometric optics [25-28]. Scatter by very small dielectric spheres, $\pi D_p < 0.2\lambda$ and $\kappa << 1$, is described by the Rayleigh formulae

$$\eta_{\lambda s} = \frac{8}{3}\left\{\frac{\pi D_p}{\lambda}\right\}^4 \left\{\frac{n^2 - 1}{n^2 + 2}\right\}^2 \tag{3.2.2}$$

$$p_\lambda = \frac{3}{4}(1 + \cos^2\theta) \, . \tag{3.2.3}$$

The particles used in marker nephelometry are, however, typically in the size range $D_p = 0.2$–$2\ \mu$m, and the operation of the system is usually in the visible region of the spectrum, $\lambda = 0.4$–$0.7\ \mu$m. Scatter by spheres in this range of

$\pi D_p/\lambda$ is described by the Mie equations which are based on a general solution of Maxwell's equations. The behaviour of $\eta_{\lambda s}$ and p_λ is here a very complicated function of λ, D_p, and m. The problem is treated in books by Van de Hulst [27] and Kirker [28], and some numerical results that are informative in the present context have been presented by Plass [29].

In the absence of interference between particles, the functions $\eta_{\lambda s}$ and p_λ are the same for each particle in a cloud as for that particle in isolation. The criterion for effectively independent scatter by monodisperse spheres is that the centre-to-centre distance between particles be over 3 radii [27,30,31]. Since this critical separation corresponds to a volume fraction of particles of about 30% (i.e. extremely high) it is evident that this criterion is certainly satisfied in marker nephelometry under any reasonable operating conditions.

Consider, then, the independent scattering of radiation in a cloud of monodisperse spherical particles. The scatter into a solid angle $\mathbf{d\Omega}_s$ from the N particles found at a given instant in a control volume element δV under incident radiation of flux density $I_\lambda d\lambda d\Omega_i$ is N times that from a single particle, thus

$$\frac{1}{4\pi} N A_p \eta_{\lambda s} p_\lambda I_\lambda \, d\lambda d\Omega_i d\Omega_s \;. \tag{3.2.4}$$

If N is large enough to constitute a statistical population, this can be written as

$$\frac{1}{4\pi} K_{\lambda s} p_\lambda I_\lambda \, d\lambda d\Omega_i d\Omega_s \, \delta V \tag{3.2.5}$$

where $K_{\lambda s} \equiv \mathcal{C} A_p \eta_{\lambda s}$ is the *spectral scatter coefficient* and $\mathcal{C}$ is the number concentration of particles. When N is large, $N = \mathcal{C}\,\delta V$.

The expression (3.2.5) is taken to be applicable to any cloud of particles, and constitutes a definition of $K_{\lambda s}$. In the case of polydisperse spherical particles of the same material, the summation of scatter over all size fractions thus gives

$$K_{\lambda s} = \mathcal{C} \int_0^\infty A_p \eta_{\lambda s} f(D_p)\, dD_p \;, \tag{3.2.6}$$

$$p_\lambda = \frac{\mathcal{C}}{K_{\lambda s}} \int_0^\infty A_p \eta'_{\lambda s} p'_\lambda f(D_p)\, dD_p \;, \tag{3.2.7}$$

where $\eta'_{\lambda s}$ and p'_λ are the scattering efficiency and phase function for particles of diameter D_p, and $f(D_p)dD_p$ is the number fraction of particles in the diameter range D_p, $D_p + dD_p$.

In the case of non-spherical particles, it can normally be assumed that the particles are randomly orientated. The scattering properties are then statistically isotropic, and expressions similar to (3.2.6) and (3.2.7) apply. For example, if the particles are all of the same shape and free of surface concavities

$$K_{\lambda s} = \mathcal{C} \int_0^\infty \frac{1}{4} S_p < \eta'_{\lambda s} > f(D_p)\, dD_p \;, \tag{3.2.8}$$

$$p_\lambda = \frac{\mathcal{C}}{K_{\lambda s}} \int_0^\infty \frac{1}{4} S_p < \eta'_{\lambda s} p'_\lambda > f(D_p)\, dD_p \;, \tag{3.2.9}$$

where S_p is the surface area of a particle of characteristic diameter D_p, $S_p/4$ is the mean projected area of randomly orientated particles of this diameter, and the angular brackets indicate appropriate averages over all orientations. In (3.2.7) as well as in (3.2.5), the directional dependence of p_λ involves only the angle θ. However, the particle size distribution is important and in general

$$p_\lambda = p_\lambda \; (\lambda, \theta, m, \text{particle shape, particle size distribution}).$$

We have now considered all the relations that express the scatter of radiation by particles in a small control volume. This scatter is the process by which the signal is produced in marker nephelometry, and it is therefore of prime interest in this technique. It is also necessary, however, to examine the transport of radiation across the field that is traversed by the beam from the light-source and by scattered radiation directed towards the optical receptor. It is here that many of the sources of error in the technique arise.

The general equation of radiative transfer for an absorbing and scattering medium is

$$\frac{\partial I_\lambda}{\partial R} + K_\lambda I_\lambda = K_\lambda \mathfrak{J}_\lambda , \tag{3.2.10}$$

where I_λ is the spectral intensity of radiation at a point $\mathbf{R}$ into a solid angle $\mathbf{d\Omega}$, $\partial I_\lambda / \partial R$ is the rate of change of I_λ with distance at $\mathbf{R}$ in the direction of $\mathbf{d\Omega}$, K_λ is the *spectral extinction coefficient*, and $\mathfrak{J}_\lambda$ is the *source function.* The quantity $K_\lambda \mathfrak{J}_\lambda$ is the rate of growth of I_λ with distance by (i) the emission of radiation into $\mathbf{d\Omega}$ by material along the considered path, and (ii) the in-scatter of radiation from other directions into $\mathbf{d\Omega}$ by particles along the same path:

$$K_\lambda \mathfrak{J}_\lambda = 4\, K_{\lambda a} E_{b\lambda} + \mathcal{G}_\lambda + \frac{K_{\lambda s}}{4\pi} \int_{4\pi} p_\lambda I'_\lambda \, d\Omega' \tag{3.2.11}$$

where $4K_{\lambda a}E_{b\lambda}$ is the effect of thermal emission, $\mathcal{G}_\lambda$ is emission by any other means, such as chemiluminescence, and the last term gives the in-scatter; I'_λ and $\mathbf{d\Omega}'$ are spectral intensities and solid angles in all directions from $\mathbf{R}$, $K_{\lambda a}$ is the *spectral absorption coefficient*, and $E_{b\lambda}$ is the *spectral emittance* given by Planck's law of blackbody emission and related to the total emittance by

$$\int_0^\infty E_{b\lambda}\, d\lambda = \sigma T^4 . \tag{3.2.12}$$

The spectral absorption coefficient $K_{\lambda a}$ is related to the concentration of absorbing particles in the same way as $K_{\lambda s}$ is related to the concentration of scatterers; e.g. for monodisperse spheres

$$K_{\lambda a} = \mathcal{C} A_p \eta_{\lambda a} , \tag{3.2.13}$$

where $\eta_{\lambda a}$ is the *efficiency factor for absorption*, and $A_{\lambda a} \equiv A_p \eta_{\lambda a}$ is the *absorption cross-section.* By definition,

$$K_\lambda = K_{\lambda a} + K_{\lambda s} . \tag{3.2.14}$$

The particles doing the absorbing and scattering may not all be the same, however. In marker nephelometry on flames, for example, the scattering particles may be the marker and soot, whereas the absorbing and thermally emitting particles also include molecules of H_2O, CO_2, and other polyatomic species. Computed results for spherical particles [28,29] show that the absorption efficiency $\eta_{\lambda a}$, like the scatter efficiency $\eta_{\lambda s}$, can take values well above unity. In general,

$$\eta_{\lambda a} \to 0 \text{ as } \pi D_p/\lambda \to 0$$

$$\eta_{\lambda a} < 1 \text{ for } \pi D_p/\lambda \ll 1$$

$$\eta_{\lambda a} < 1 \text{ as } \pi D_p/\lambda \to \infty .$$

The sum $\eta_\lambda \equiv \eta_{\lambda s} + \eta_{\lambda a}$ is the extinction efficiency. It has the property [27]

$$\eta_\lambda \to 2 \text{ as } \pi D_p/\lambda \to \infty .$$

Integration of (3.2.10) along a line L in the given direction through **R** gives

$$I_\lambda = I_{\lambda o}\tau_{\lambda L} + \tau_{\lambda L}\int_0^L K_\lambda \mathfrak{J}_\lambda \tau_\lambda^{-1}\, dR \quad , \tag{3.2.15}$$

where

$$\tau_\lambda \equiv \exp\left\{ -\int_0^R K_{\lambda s}\, dR \right\} \tag{3.2.16}$$

is the *spectral transmissivity* for radiation travelling the path from $R = 0$ up to any point between $R = 0$ and $R = L$, i.e., it is the fraction of such radiation that reaches the given point. The transmissivity of the whole path is

$$\tau_{\lambda L} \equiv \exp\left\{ -\int_0^L K_{\lambda s}\, dR \right\} . \tag{3.2.17}$$

3.3 *Optical Output of a Nephelometer Probe*

The total radiant flux emitted in all directions by the radiating matter in a light source (lamp or laser) is $\eta_L P_L$, where η_L is the radiant efficiency, and P_L is the electrical power input. The spectral flux sent by the source – through an optically clear ambient medium – into a solid angle $\mathbf{d}\Omega$ at a distance so great that the source is virtually a point is $J_{\lambda L}\, d\Omega$, with

$$\eta_L P_L = \int_{2\pi}\int_0^\infty (J_{\lambda L}/T_{\lambda L})\, d\lambda\, d\Omega \tag{3.3.1}$$

where $J_{\lambda L}$ is the effective *spectral point-source intensity*, and $T_{\lambda L}$ is the effective *transmittance* of the source window (bulb or tube) in a given direction. The total flux leaving the source in a solid angle $\mathbf{d}\Omega$ is $J_L\, d\Omega$,

$$J_L = \int_0^{\infty} J_{\lambda L}\, d\lambda \;, \tag{3.3.2}$$

and the total flux in all directions is

$$\Phi_L = \int_{2\pi} J_L\, d\Omega \;. \tag{3.3.3}$$

Lamps (arc, tungsten, gas-discharge) are usually rated in terms of the input power P_L. Lasers, on the other hand, are usually rated in terms of the total output radiant flux Φ_L. The directional characteristics of both lamps and lasers are typically described in terms of J_L.

Whatever the exciter light source may be, the source flux of primary interest in marker nephelometry is the flux $\Phi_{\lambda E} \equiv \Phi_{\lambda L}\eta_E$ that is directed into the aperture of the exciter optics, where η_E is the collection efficiency. The exciter optics (if any) conduct, filter, split, bend, collimate, and/or focus the light. The spectral flux put out in the probe's incident beam is then $\Phi_{\lambda E} T_{\lambda E}$ where $T_{\lambda E}$ is the overall spectral transmittance of the exciter optics. The collection efficiency η_E is usually less than 20% for an arc lamp, whereas for a laser it can be unity. The transmittance $T_{\lambda E}$ is unity for a laser with no exciter optics. At the other extreme, transmission through a lengthy fibre-optic conduit may give values of as small as 0.01.

It will be taken, unless otherwise stated, that (i) the function of the marker nephelometer probe is to detect the marker concentration at a point in a field, (ii) the probe's incident beam for this application is axisymmetrical, and (iii) the region of the incident beam from which scattered light is taken by the detector – the probe control volume (pcv) – is defined by a slit in the detector optics, as indicated in Fig.1. The length of the line segment along the incident beam axis between the ends of the pcv is L_i, and its centre O_i is taken to be the centre of the pcv. The optical (viewing) axis of the detector – the axis of the conical solid angle in which light scattered from the pcv is gathered – can usually be defined as the line joining the pcv centre O_i and its image point O_i' in the plane of the slit. The effective scattering angle θ_s is taken as the angle between the vector $\overrightarrow{O_iO_i'}$ and the vector $\boldsymbol{\delta}_i$, the latter being a unit vector along the axis of the incident beam in the direction of the incident light flux. It will be assumed that the ends of the pcv are effectively parallel planes a distance $L_i \sin\theta_s$ apart, and that variations in the radial energy distribution of the incident beam with distance along its axis are negligible within the pcv. A pcv thus defined with $\theta_s = 90°$ will be called a *right-cylindrical pcv*, and one with any other θ_s an *oblique-cylindrical pcv*. The length L_i – the optical depth of the pcv with respect to the incident beam – will be called the *incident optical depth* of the pcv.

Consider, then, a control plane or screen S_i – the probe's *plane of normal incidence* – that passes through the centre of the pcv and is orientated normal to the axis of the incident beam. The spectral flux density upon a screen element dS_i at radius r from the axis is $G_{\lambda i}(r)$. It can generally be assumed that if the operating conditions are within the range of feasibility for marker nephelometry, then the effects of emission and in-scatter on the radiant flux within the solid angle of the incident beam are quite negligible. Thus

$$\int_0^\infty 2\pi r G_{\lambda i} dr = \Phi_{\lambda E} T_{\lambda E} T_{\lambda i} \quad , \tag{3.3.4}$$

where $T_{\lambda i}$ is the transmittance of the path along the incident beam up to the pcv. If this path were optically clear, the flux density on dS_i would be $G_{\lambda i}(r) = \Phi_{\lambda E} T_{\lambda E} \varphi_i(r)$, where $\varphi_i(r)$ is a function describing the radial distribution of flux density in the full incident beam on S_i. Equation (3.3.3) thus shows that in the presence of the optically active field

$$G_{\lambda i}(r) = \Phi_{\lambda E} T_{\lambda E} T_{\lambda i} \varphi_i(r) \ . \tag{3.3.5}$$

It was supposed in earlier discussions [19,20] that the incident beam is sharp-edged and uniform, of diameter D_i, so that φ_i is the step function

$$\varphi_i(r) = \begin{cases} 4/\pi D_i^2 , & r < D_i/2 \\ 0, & r > D_i/2 \ . \end{cases} \tag{3.3.6}$$

This is usually a reasonable approach for systems in which an arc lamp is used with a combination of lenses and a diaphragm with a circular aperture to define D_i, and D_i is large compared to the Airy disk diameter d_A, crudely $D_i >> \lambda$. The present approach provides the generality needed to deal with a non-uniform radial distribution of beam energy; e.g., the "Gaussian" distribution that is typical for laser beams,

$$\varphi_i(r) = \frac{2}{\pi a_i^2} e^{-2r^2/a_i^2} \quad , \tag{3.3.7}$$

where $\varphi_i(a_i)/\varphi_i(0) = 1/e^2$.

It may be desirable, in characterizing the pcv geometry, to define an effective beam diameter D_i for a beam with an energy distribution such as (3.3.7). The following definition is suggested:

$$\int_0^{b_i} 2\pi r \varphi_i(r)\, dr = ½ \, , \qquad D_i \equiv 2b_i\sqrt{2} \ , \tag{3.3.8}$$

where b_i is the radius of the circle containing half the energy flux, and D_i is the true beam diameter in the case of the uniform distribution (3.3.5). This diameter can be used, for example, in characterizing the volume of the pcv, $V = (\pi/4)D_i^2 L_i$; the average radiant flux density in the pcv, $G_{\lambda i} = \Phi_{\lambda E} T_{\lambda E}/(\pi D_i^2/4)$; and the effective diameter of the pcv in respect to volume-averaging – discussed in Sec.3.5, $D_V = (6V/\pi)^{1/3}$. It is shown in Sec.3.6 that the effective diameter of an incident beam with the energy distribution (3.3.7) in respect to marker shot noise is $2a_i$.

We shall suppose, in evaluating the light scattered into the detector aperture, that the solid angle Ω_i of the incident beam, and the solid angle Ω_s of the scattered beam as defined by the detector aperture, are both quite small. The spectral flux scattered by marker particles from the pcv into Ω_s can then be estimated from (3.1.5) without integration over Ω_i and Ω_s. We substitute Ω_s for $d\Omega_s$ and $G_{\lambda i}$ for $I_\lambda d\Omega_i$ and obtain

$$\frac{1}{4\pi}\Omega_s p^*_{\lambda s}\int_{pcv} K^*_{\lambda s}G_{\lambda i}\,dV \;, \tag{3.3.8}$$

where $p^*_{\lambda s} \equiv p^*_\lambda(\theta_s)$ and the asterisk denotes a property of the marker particles. The fraction of this flux transmitted along the path between the pcv and the detector is $T_{\lambda s}$, and the fraction that is then transmitted by the detector optics to the phototransducer is $T_{\lambda D}$. Thus the total marker-scattered flux from the pcv reaching the phototransducer is

$$\Phi^*_{\lambda s} = \frac{1}{4\pi}\Omega_s p^*_{\lambda s}T_{\lambda s}T_{\lambda D}\int_{pcv} K^*_{\lambda s}G_{\lambda i}dV \;. \tag{3.3.9}$$

If the pcv is small enough so that variation of the marker concentration (and hence of $K^*_{\lambda s}$) is negligible within it, then (3.3.4) and (3.2.9) together give

$$\Phi^*_{\lambda s} = \frac{1}{4\pi}\Omega_s\Phi_{\lambda E}T_{\lambda N}T_{\lambda W}\tau_\lambda p^*_{\lambda s}K^*_{\lambda s}L_i \;, \tag{3.3.10}$$

where $T_{\lambda N} \equiv T_{\lambda E}T_{\lambda D}$ is the total transmittance of the probe (nephelometer) optical components, $T_{\lambda W}\tau_\lambda \equiv T_{\lambda i}T_{\lambda s}$, $T_{\lambda W}$ is the total transmittance of windows or optically transmitting walls in vessels or conduits containing the fluid that is being probed, and τ_λ is the total transmissivity of the field traversed in that fluid by the incident and scattered beams. The value of τ_λ for small Ω_i and Ω_s is given by

$$\tau_\lambda = \exp\left(-\int K_\lambda\, dR\right) \;, \tag{3.3.11}$$

where the integral is over the path along the incident beam to the pcv, and along scattered beam from the pcv, and $K_\lambda = K_{\lambda a} + K_{\lambda s}$ is the total extinction coefficient. If there are no scattering particles other than the marker, then $K_{\lambda s} = K^*_{\lambda s}$.

In view of the usual objectives of marker nephelometry, it is desirable to express the above results in terms of the marker concentration. We shall therefore substitute

$$K^*_{\lambda s} \equiv k^*_{\lambda s}\Gamma^* \;, \tag{3.3.12}$$

where Γ^* is the marker mass concentration. The specific scatter coefficient $k^*_{\lambda s}$ is, under the usual condition of independent scatter by all scattering particles, independent of Γ^*.

It is evident that (3.3.10) suggests an ideal response which is never quite realized in practice. The total spectral flux actually impinging on the phototransducer is

$$\Phi_{\lambda P} = \Phi^*_{\lambda s} + \Phi_{\lambda b} \;. \tag{3.3.13}$$

where $\Phi^*_{\lambda s}$ is the flux due to the scatter of light from the incident beam by marker particles in the pcv and $\Phi_{\lambda b}$ is the background flux due to all other effects. The marker-scattered light flux can be resolved as

$$\Phi^*_{\lambda s} = \Phi^*_\lambda + \Phi'_{\lambda V} + \Phi'_{\lambda msn} - \Phi'_{\lambda \tau} \ , \qquad (3.3.14)$$

the components being:

1. The basic signal

$$\Phi^*_\lambda \approx \frac{1}{4\pi} \Omega_s \Phi_{\lambda E} T_{\lambda N} T_{\lambda W} p^*_{\lambda s} k^*_{\lambda s} L_i \Gamma^* \qquad (3.3.15)$$

where Γ^* is the marker concentration at the centre of the pcv, and Φ^*_λ is the signal that would be obtained if the marker concentration were uniformly Γ^* throughout the pcv, if $\tau_\lambda = 1$, and if there were no *marker shot noise.*

2. The distortion $\Phi'_{\lambda V}$ due to spatial averaging of Γ^* when Γ^* is not effectively uniform over the pcv. This effect is discussed in Sec.3.4.

3. The distortion $\Phi'_{\lambda msn}$, the marker shot noise, which becomes prominent when the number of marker particles in the pcv is not large enough to constitute a statistical population. This effect is discussed in Sec.3.12.

4. The distortion in the above three effects due to optical extinction along the probe light path through the field,

$$\Phi'_{\lambda \tau} = (\Phi^*_\lambda + \Phi'_{\lambda V} + \Phi'_{msn})(1 - \tau_\lambda) \ . \qquad (3.3.16)$$

It may be asked, incidentally, how Γ^* can be defined at all when the number of particles in the pcv is small. The answer is that it is always possible to imagine a better marker with the same scattering properties and particle diffusivity which provides an adequate population for the given conditions.

The optical background flux $\Phi_{\lambda b}$ can also be resolved into components:

$$\Phi_{\lambda b} = \Phi_{\lambda o} + \Phi_{\lambda l} + \Phi_{\lambda m} + \Phi_{\lambda lm} \ . \qquad (3.3.17)$$

The terms of this expression represent the following effects:

$\Phi_{\lambda o}$ = effect with the probe light source and marker particle source both off, normally the in-scatter of room light by naturally occurring particles and the reflection of room light from surfaces; in flames and similar hot or reacting systems it also includes thermal emission and chemiluminescence.

$\Phi_{\lambda l}$ = effect superimposed on $\Phi_{\lambda o}$ when the light source is on, normally the scatter of source light from the pcv and the in-scatter of source light from elsewhere in the field due to naturally occurring particles, and the reflection of scattered or reflected source light by surfaces.

$\Phi_{\lambda m}$ = effect superimposed on $\Phi_{\lambda o}$ when the marker source is on, normally the in-scatter of room light by the marker; in flames it also includes thermal emission by the marker.

$\Phi_{\lambda lm}$ = effect superimposed on $\Phi_{\lambda o}$, $\Phi_{\lambda m}$ and $\Phi_{\lambda l}$ when the light source and the marker source are both on, normally the in-scatter of source light by the marker and the reflection of marker-scattered source light by surfaces.

Some aspects of these effects can, in principle, be estimated by solving the equation of transfer, (3.2.10). All of the effects can, however, be studied experimentally, and this is normally the only useful approach. The experimental observation of the fourth effect requires that the pcv be blocked from view of the receptor. It is expected that these effects should, on the whole, be roughly proportional to Ω_s.

3.3 *Electrical Output of a Nephelometer Probe*

Suppose the phototransducer receives as input a radiant flux

$$\Phi_P = \int_0^\infty \Phi_{\lambda P}\, d\lambda \tag{3.4.1}$$

which is steady but for the fluctuations – or *photonic shot noise* – due to the particulate nature of radiation. The resulting output is ideally an electric current I which is steady but for the fluctuations caused by the photonic shot noise and by random variations in the photoemissive yield of electrons per photon. The quantum efficiencies of the photosensitive elements (photocathodes) of photoelectric transducers are generally less than unity; thus the random train of photons yields a random train of electrons which does not differ distinguishably from any other randomly generated train of electrons. The fluctuations in the current I can therefore be regarded as simply reflecting the particulate nature of electricity, and constitute what is called *electronic shot noise.* The relation between the time-mean values of radiant flux and current is generally written as

$$\bar{I} = \int_0^\infty s_{\lambda P}\bar{\Phi}_{\lambda P}\, d\lambda + \bar{I}_d \ , \tag{3.4.2}$$

where $s_{\lambda P}$ is the spectral sensitivity of the phototransducer and $\bar{I}_d$ is the mean *dark current* – the background current or bias due to electron emission by means other than photoemission (primarily thermoemission). Data on $s_{\lambda P}$ and $\bar{I}_d$ are usually included in manufacturers' literature. For the phototransducers (photomultiplier tubes and photodiodes) that are most suitable for marker nephelometry, $s_{\lambda P}$ is generally independent of the light flux, and $\bar{I}_d$ can be assumed to be the same under any conditions of photoemission as in darkness.

Shot noise is ordinarily of no concern in the operation of instruments, and ordinarily there is no need for time-averaging in a relation such as (3.4.2) for "steady" conditions to make it understood that the direct-current response is meant. In the operation of phototransducers at low light levels, however, the fluxes of photons and electrons can be low enough to bring such fluctuation noise into prominence. Since shot noise is purely fluctuation noise, it only affects the detection of marker concentration *flucutations* in marker nephelometry. The mean marker concentration can, in principle, be detected at any level of such noise; however, difficulty is eventually encountered because of excessive requirements for time-averaging, and by masking of the signal by other forms of noise such as the dark current.

Consider, then, a nephelometer probe operating on a statistically stationary turbulent field. Equation (3.4.2) applies to the mean output current $\overline{I}$, which can be resolved as

$$\overline{I} = \overline{I}_s^* + \overline{I}_b + \overline{I}_d \ , \tag{3.4.3}$$

where

$$\overline{I}_s^* = \int_0^\infty s_{\lambda p} \overline{\Phi}_{\lambda s}^* \, d\lambda \tag{3.4.4}$$

$$\overline{I}_b = \int_0^\infty s_{\lambda p} \overline{\Phi}_{\lambda b} \, d\lambda \ . \tag{3.4.5}$$

The response $\overline{I}_s^*$ to marker-scattered light can be further resolved as indicated by (3.3.14). The result is conveniently expressed as

$$\overline{I}_s^* = P \overline{\tau} \overline{I}^* \ , \tag{3.4.6}$$

where $\overline{\tau}$ is the mean optical transmissivity over the paths of the incident and scattered beams across the field under study,

$$\overline{I}^* \equiv \int_0^\infty s_{\lambda p} \overline{\Phi}_\lambda^* \, d\lambda = s^* \overline{\Gamma}^* \ , \tag{3.4.7}$$

$$P \equiv \overline{\Phi}_\lambda^* / (\overline{\Phi}_\lambda^* + \overline{\Phi}_{\lambda V}') \ , \tag{3.4.8}$$

and s^* is the output current sensitivity to the mean marker concentration. It is assumed in (3.4.6) that the covariance of fluctuations in τ and I^* is negligible relative to $\overline{\tau}\overline{I}^*$. The effect of shot noise on $\overline{I}_s^*$ is nil. The volume-averaging function P is discussed in Sec.3.5.

The response $\overline{I}_b$ to the background light flux can similarly be resolved as indicated by (3.3.17):

$$\overline{I}_b = \overline{I}_o + \overline{I}_m + \overline{I}_l + \overline{I}_{ml} \ . \tag{3.4.9}$$

The terms in this expression must normally be investigated empirically, as suggested by the discussion below (3.3.17).

The fluctuations in the phototransducer output current, $i = I - \overline{I}$, are more difficult to interpret than the mean response $\overline{I}$. The analogue of (3.4.3) for the mean-square fluctuation in a frequency interval f, $f + \Delta f$ is

$$\overline{i^2}\big|_{f,\Delta f} = \left(\overline{i_s^{*2}} + 2\overline{i_s^* i_b} + \overline{i_b^2} + \overline{i_{esn}^2}\right)\Big|_{f,\Delta f} \ , \tag{3.4.10}$$

where i_s^* and i_b may be correlated, but the electronic shot noise current i_{esn} and i_s^* or i_b are not. It is assumed, as is usually the case, that fluctuations in the dark current due to factors other than electronic shot noise are negligible. The term in i_{esn} in (3.4.10) is the total electronic shot noise from all components of current — I_s^*, I_b and I_d.

The first term in parentheses in (3.4.10) can be written as

$$\overline{i_s^{*2}}\Big|_{f,\Delta f} \;=\; \left(\bar{\tau}^2 Q\overline{i^{*2}} + \bar{\tau}^2\overline{i^2_{msn}} + \overline{i^2_E} + \ldots\right)\Big|_{f,f+\Delta f} \qquad (3.4.11)$$

where

$$\overline{i^{*2}}\Big|_{f,\Delta f} \;=\; s^{*2}\,\overline{\gamma^{*2}}\Big|_{f,\Delta f}\,, \qquad (3.4.12)$$

$\gamma^* \equiv \Gamma^* - \bar{\Gamma}^*$ is the marker concentration fluctuation, i_{msn} is the marker shot noise current, and i_E is the effect of fluctuations in the radiant energy output of the exciter light source. The volume-averaging function Q is discussed in Sec.3.5. It is assumed that the current sensitivity s^* for γ^* is the same as that in (3.4.7) for $\bar{\Gamma}^*$; this requires that the phototransducer frequency response be flat over the turbulence range – a condition that is usually easy to satisfy. Not shown in (3.4.11) are a number of terms that are relatively small under reasonable operating conditions – principally those in $\gamma^*\acute{\tau}$, where $\acute{\tau}$ is the fluctuation in τ.

The covariance term $\overline{i_s i_b}$ is awkward because there is normally no easy way of measuring or predicting it directly. It should be made as small as possible by ensuring that the background radiant flux Φ_b is no bigger than necessary, but the amount of control that can be exercised is unfortunately negligible in some systems, e.g. flames. Its maximum magnitude can be ascertained from

$$\mathrm{abs}(\overline{i_s^*\, i_b})\Big|_{f,\Delta f} < \sqrt{(\overline{i_s^{*2}}\;\overline{i_b^{\,2}})}\Big|_{f,\Delta f}\,. \qquad (3.4.13)$$

In the event of a significant mass concentration Γ^+ of naturally-present scattering particles, a major component of this covariance is a term in

$$\overline{i^* i^+}\Big|_{f,\Delta f} \;=\; s^* s^+\,\overline{\gamma^*\gamma^+}\Big|_{f,\Delta f} \qquad (3.4.14)$$

where $\gamma^+ \equiv \Gamma^+ - \bar{\Gamma}^+$ is the fluctuation in Γ^+, s^+ is the current sensitivity to $\bar{\Gamma}^+$ defined as in (3.4.7), and i^+ is the current response to γ^+. A discussion of $\overline{\gamma^*\gamma^+}$ is beyond the present scope; it is a question that must be considered, however, in any application of marker nephelometry to systems such as sooty flames or dust-laden jets. A suggested expression for $\overline{i_s i_b}\big|_{f,\Delta f}$ that incorporates (3.4.14) is

$$\overline{i_s^* i_b}\Big|_{f,\Delta f} \;=\; \left(\bar{\tau}^2 Q\overline{i^* i^+} + \ldots\right)\Big|_{f,\Delta f} \qquad (3.4.15)$$

where $\bar{\tau}$ and Q are assumed to be the same as in (3.4.11).

The frequency spectral density function of the current fluctuations is defined by

$$H(f) \;\equiv\; \lim\left\{\frac{\overline{i^2}\big|_{f,\Delta f}}{\Delta f}\right\}, \quad \Delta f \to 0\;; \qquad (3.4.16)$$

$$\overline{i^2}\big|_{f,\Delta f} \equiv \int_f^{f+\Delta f} H(f)\, df \, . \tag{3.4.17}$$

Thus (3.4.11) is equivalent to

$$H_s^*(f) = \overline{\tau}^2 Q_f H^*(f) + \overline{\tau}^2 H_{msn}(f) + H_E(f) + \ldots, \tag{3.4.18}$$

where

$$H^*(f) = s^{*2} G^*(f) \ , \tag{3.4.19}$$

$$\overline{\gamma^{*2}}\big|_{f,\Delta f} = \int_f^{f+\Delta f} G^*(f)\, df \ , \tag{3.4.20}$$

and $G^*(f)$ is the spectral density function of the marker concentration fluctuations.

3.5 Volume Averaging

Becker, Hottel and Williams [20] have shown that the effect of volume-averaging on the resolution of the mean marker concentration is negligble when the mean concentration gradient is virtually uniform within the pcv. This condition is usually easy to satisfy; then $P \doteq 1$ in (3.4.6).

The resolution of $\overline{\gamma^{*2}}\big|_{f,\Delta f}$ was also investigated theoretically and experimentally, and the results of that study will now be summarized. It was supposed that $\overline{\Gamma}^*$ and $\overline{\gamma^{*2}}$ are effectively uniform throughout the pcv — a mathematically reasonable simplification. The volume-averaging function $Q_{f,\Delta f}$ in (3.4.11) defines

$$\overline{\breve{\gamma}^{*2}}\big|_{f,\Delta f} = \overline{\gamma^{*2}}\big|_{f,\Delta f}\, Q_{f,\Delta f} \ , \tag{3.5.1}$$

where $\breve{\gamma}^*$ is the volume-averaged concentration flucutation. In the case of a uniform incident lightbeam giving a sharply-defined pcv of volume $V = \pi D_i^2 L_i/4$,

$$\breve{\gamma}^* = \int_V \gamma^*\, dV \, . \tag{3.5.2}$$

The frequency spectral density function of γ^* is defined by (3.4.20); the corresponding function for $\breve{\gamma}^*$ is

$$G_V^*(f) = Q_f G^*(f) \ , \tag{3.5.3}$$

where Q_f is the same factor that appears in (3.4.11). Suppose that the magnitude of the mean velocity vector, $\overline{U} \equiv (\overline{U}_x^2 + \overline{U}_y^2 + \overline{U}_z^2)^{1/2}$, is large compared to the velocity fluctuation intensity, so that $G^*(f)$ reflects the convection of a virtually frozen concentration field past the point of measurement; then

$$G^*(f) \doteq \frac{2\pi}{\overline{U}}\, E^*(\kappa) \ , \tag{3.5.4}$$

where $E^*(\kappa)$ is the one-dimensional wave-number spectral density function of

γ^*, and $\kappa = 2\pi f / \bar{U}$ is the wave number component in the direction of mean motion. Then

$$E_V^*(\kappa) \;=\; Q_\kappa E^*(\kappa) \;. \tag{3.5.5}$$

The theory of Becker *et al.* [20] for $\kappa \to 0$ gives

$$Q_\kappa \;=\; 1 - 0.16\,\kappa^2 D_V^2 \tag{3.5.6}$$

for right-cylindrical pcv's with $L_i = 0.8\,D_i$, where $D_V = (6V/\pi)^{1/3}$ is the diameter of a sphere of equal volume. The advantage of using D_V in (3.5.6) is that the numerical coefficient is insensitive to moderate variations in L_i/D_i and θ_s [20]. The experimental data of $G_V(f)$ in a turbulent free jet by the same authors were well fitted by:

$$Q_\kappa \;=\; \exp(-0.12\,\kappa^2 D_V^2) \;. \tag{3.5.7}$$

This agrees reasonably well with (3.5.6) for $\kappa D_V < 1$. It may be concluded, subject to the conditions noted below, that:

1. The correction on the spectral density is of the general form

$$Q_\kappa \;=\; Q_\kappa(\kappa D,\; L_i/D_i,\; \varphi_i(r/D_i),\; \theta_s) \tag{3.5.8}$$

where D is any characteristic dimension of the pcv, such as L_i, D_i, and D_V.

2. Equation (3.5.7) is generally valid for spectrum measurements with any nephelometer probe whose incident beam is virtually uniform and whose pcv is approximately right-cylindrical with $L_i/D_i = 0.6 - 1.1$.

3. The behaviour of Q_κ for pcv's with other geometries (different values of θ_s and φ_i) should be similar to (3.5.7). Thus measurements at a few well-chosen frequencies f with two or three geometrically similar pcv's having $D_2^2 = 2D_1^2$, $D_3^2 = 3D_1^2$, ... should suffice to determine Q_κ.

The conditions on these conclusions are:

1. The theory of Becker *et al.* [20] is limited to regions of the spectrum where the turbulence is roughly isotropic, say $\Lambda_\gamma \kappa > 1$, where Λ_γ is the integral scale of the field of γ^* in the direction of the mean velocity $\bar{\mathrm{U}}$. The experimental results obtained were generally such that $Q_\kappa \doteq 1$ for $\Lambda_\gamma \kappa < 2$. From a practical viewpoint, when marker nephelometry is used to measure concentration fluctuations the pcv's should at least be small enough so that $Q_\kappa \approx 1$ up to $\Lambda_\gamma \kappa = 2$, otherwise the corrections will be so large over a major part of the spectrum that the validity of the results may be highly questionable.

2. The theory does not take into account the effect of intermittency, and the experimental results were for a non-intermittent region. In the case of systems such as a marked turbulent jet in a laminar free stream, a reasonable correction for this effect can perhaps be made by assuming that the pcv is small enough so that, to a good approximation, the pcv is always either inside the turbulent fluid or out. The effect of intermittency should be to shift the region of validity of the theory

to higher values of $\Lambda_\gamma \kappa$ where the fluctuation energy is primarily due to the turbulent fluid; the problem warrants experimental study. In the author's applications of marker nephelometry he has always worked with pcv's small enough for corrections for volume-averaging to be small, and the possibility of small errors in intermittent regions was not a serious consideration.

The effect of volume-averaging on the mean-square concentration fluctuation in a frequency interval $f, f+\Delta f$ can be evaluated from the spectrum:

$$\overline{\breve{\gamma}^{*2}}\Big|_{f,\Delta f} = \int_f^{f+\Delta f} G_V(f)\, df = \int_f^{f+\Delta f} Q_f G(f)\, df = Q_{f,\Delta f}\, \overline{\gamma^{*2}}\Big|_{f,\Delta f} . \qquad (3.5.9)$$

Some further remarks on this problem are given in Sec.4.2.

It may be noted that in the case of a right-cylindrical pcv and an incident beam with the energy distribution function (3.3.7), the volume-average value of γ^* is

$$\breve{\gamma}^* = \int_0^{L_i}\int_0^{2\pi}\int_0^{\infty} 2\pi r^2\, \varphi_i(r)\, \gamma^*\, dr\, d\phi\, dx .$$

The modification required for oblique pcv's, $\theta_s \neq 90°$, is obvious. The theory of Becker *et al.* [20] for Q_f also applies to these pcv's, and predictions of Q_f for $\kappa D_V << 1$ could be made accordingly. It appears though, from what has already been said, that such additional theoretical guidance is not particularly needed.

3.6 Marker Shot Noise

Suppose that (i) the marker particles are monodisperse spheres, (ii) the incident light beam is uniform, and (iii) the pcv is sufficiently small that $Q_f \approx 1$ over the frequency range $0 < f < \Delta f$ that contains the bulk of the marker shot noise. It has been shown [19,20] that in this case

$$\overline{(N - N^*)^2} = \bar{N} = \bar{N}^* \qquad (3.6.1)$$

where N is the number of marker particles in the pcv at time t, m_p is the mass of a particle, and $N^* \equiv V\Gamma^*/m_p$ is the number required to represent the instantaneous marker concentration (concentration being defined, as always, in terms of statistical populations). The mean-square fluctuation in the output current of the phototransducer that is due to marker shot noise is therefore

$$\overline{i^2_{msn}} = \bar{I}^{*2}/\bar{N}^* = \bar{I}^{*2} m_p/V\bar{\Gamma}^* . \qquad (3.6.2)$$

Becker *et al.* [20] gave an order-of-magnitude representation of the spectral density function $H_{msn}(f)$ of the current fluctuations i_{msn}. A more accurate estimate is, however, easily derived. Consider the $n(t) \equiv N(t) - N^*(t)$ "noisy" particles that are in the pcv at a given time. At a certain time $t = t'$ the number of such particles is $n' \equiv n(t')$, while at any later instant it is $n(t) \equiv n'' + \Delta n$, where $n'' \equiv n(t', t)$ is the remainder of those particles that were present at $t = t'$ and Δn is the number of new noisy particles. From the nature of the process, no correlation is expected between Δn and $n' \equiv n(t')$. Thus

$$\overline{n(t')\,n(t)} = \overline{n(t')\,n(t',t)} \equiv \overline{n'n''} \ . \tag{3.6.3}$$

To evaluate the autocovariance $\overline{n'n''}$, consider first its average value $\langle (n'n'')_{n'} \rangle$ for a given value of n'. Suppose that particles crossing the pcv are, in effect, convected in a frozen pattern at the mean flow velocity $\overline{\mathbf{U}}$. Then

$$\langle (n'n'')_{n'} \rangle = \begin{cases} n' \langle (n'')_{n'} \rangle , & 0 \leqslant \tau \leqslant \tau_m \\ 0, & \tau > \tau_m \end{cases} \tag{3.6.4}$$

where $\tau \equiv t - t'$, and τ_m is the maximum residence time of particles in the pcv. Now

$$\langle (n'')_{n'} \rangle / n' = F(\tau) \ , \tag{3.6.5}$$

where $F(\tau)$ is the statistical residence-time distribution function of particles in the pcv, i.e., it is the ensemble-average fraction of particles that stay in the pcv for a time longer than τ. It follows from this and (3.6.4) that

$$\overline{n'n''} = \int_{-\infty}^{+\infty} \langle (n'n'')_{n'} \rangle \, f(n')\, dn' = \overline{n^2}\, F(\tau) = \bar{N} F(\tau) \ , \tag{3.6.6}$$

where $f(n)$ is the probability density function of n.

The autocovariance of the fluctuations in the phototransducer output current due to marker shot noise is thus seen to be

$$\overline{i_{msn}(t)\, i_{msn}(t+\tau)} = \overline{i_{msn}(t)\, i_{msn}(t-\tau)} = \overline{i^2_{msn}}\, F(\tau) \tag{3.6.7}$$

for $0 \leqslant \tau \leqslant \tau_m$, and zero for greater values of τ, where $\tau_m = L_m / \bar{U}$ is the maximum residence time in the pcv, L_m being the maximum particle pathlength. The corresponding autocorrelation function is

$$R_{msn}(\tau) = \begin{cases} F(\tau) , & 0 \leqslant \tau \leqslant \tau_m \\ 0, & \tau > \tau_m \ . \end{cases} \tag{3.6.8}$$

The frequency spectral density function $H_{msn}(f)$ is related to R_{msn} by

$$H_{msn}(f) = 4\, \overline{i^2_{msn}} \int_0^{\infty} R_{msn}(\tau) \cos(2\pi f \tau)\, d\tau \ . \tag{3.6.9}$$

Thus the frequency spectrum of i_{msn} can be calculated for any given residence-time distribution $F(\tau)$. The combination of (3.6.8) and (3.6.9) shows that the results are of the general form

$$H_{msn}(f) = \frac{4\overline{i^2_{msn}}\, L_m}{\bar{U}} \int_0^\infty \theta_m^{-1}\, F(\theta/\theta_m) \cos\theta \, d\theta \tag{3.6.10}$$

where $\theta \equiv 2\pi f\tau$ and $\theta_m \equiv 2\pi f L_m/\bar{U}$. Two cases are of particular interest:

1. The velocity vector $\bar{\mathbf{U}}$ is parallel to the incident beam. Then all particles have the same residence time $\tau_r = \tau_m = L_i/\bar{U}$, and

$$F(\tau) = \begin{cases} 1 - \tau/\tau_m\,, & 0 \leqslant \tau \leqslant \tau_m \\ 0, \quad \tau > \tau_m\,. \end{cases} \tag{3.6.11}$$

The corresponding spectral density function is

$$H_{msn}(f) = \frac{\overline{i^2_{msn}}\, \bar{U}}{\pi^2 f^2 L_i} \left\{ 1 - \frac{2\pi f L_i}{\bar{U}} \right\} . \tag{3.6.12}$$

2. The pcv is a right-circular cylinder of diameter D_i with its axis normal to $\bar{\mathbf{U}}$. Then

$$F(\tau) = A'(\tau)/A = 1 - \frac{2}{\pi}\frac{\tau}{\tau_m}\sqrt{1 - \frac{\tau^2}{\tau_m^2}} - \frac{2}{\pi}\sin^{-1}\frac{\tau}{\tau_m} \tag{3.6.13}$$

for $0 \leqslant \tau \leqslant \tau_m$, and $F(\tau) = 0$ for $\tau \geqslant \tau_m$, where $\tau_m = D_i/\bar{U}$, $A = \pi D_i^2/4$, and $A'(\tau)$ is the overlapping area or intersection of two circles of diameter D whose centres are a distance $D_i\tau/\tau_m$ apart. Physically, $A'(\tau)/A = V'(\tau)/V$, where V is the fixed (Eulerian) pcv, V^* is the convected (Lagrangian) control volume containing the particles found in the pcv at $\tau = 0$, and $V'(\tau)$ is the common region of V and V^*. Here (3.6.9) is not analytically integrable. Equation (3.6.13) also applies to the case of an oblique-cylindrical pcv with its ends parallel to, and its axis normal to, $\bar{\mathbf{U}}$.

The above theory does not account for the effects of factors such as volume-averaging, non-uniformity of the incident beam, and dispersion in the particle size distribution.

The effect of volume averaging in the case of a uniform incident beam and uniform particles is simple; (3.6.1) and (3.6.2) become

$$\overline{(N - \breve{N}^*)^2} = \breve{N}^* , \tag{3.6.14}$$

$$\overline{i^2_{msn}} = P^2 \bar{I}^{*2}/\breve{N}^* = P\bar{I}^{*2} m_p / V\bar{\Gamma}^* , \tag{3.6.15}$$

where

$$\breve{N}^* = \frac{1}{m_p}\int_V \Gamma^* \, dV .$$

Since $P \doteq 1$ is normally satisfied, the correction for volume averaging is rarely significant.

The effect of a laser-type of energy distribution in the incident beam, (3.3.7), on $\overline{i^2_{msn}}$ in the case of uniform particles and negligible volume-averaging is basically indicated by the combination of (3.6.2) and (3.4.7),

$$\overline{i^2_{msn}} = \bar{I}^{*2}/\bar{N}^* = \bar{I}^{*2} m_p / V\bar{\Gamma}^* = s^{*2} m_p V\bar{\Gamma}^* , \tag{3.6.16}$$

which shows that the contributions from different control volume elements are additive. A quantity I'' is defined by

$$\bar{I}^* = 2\pi \int_0^\infty I'' r dr . \tag{3.6.17}$$

Then

$$\overline{i^2_{msn}}\Big|dA = \frac{(I''dA)^2}{(\bar{\Gamma}^*/m_p)L_i dA} = \frac{2\pi m_p I''^2 r dr}{\bar{\Gamma}^* L_i} \tag{3.6.18}$$

where $dA \equiv 2\pi r dr$. From (3.3.7)

$$I'' = \frac{2\bar{I}^*}{\pi a_i^2} e^{-2r^2/a_i^2} \tag{3.6.19}$$

It follows that

$$\overline{i^2_{msn}} = 2\pi \int_0^\infty (m_p \bar{I}''^2/\bar{\Gamma}^* L_i)\, r dr = \frac{m_p \bar{I}^{*2}}{\pi a_i^2 L_i \bar{\Gamma}^*} \equiv \frac{\bar{I}^*}{\bar{N}^*_{eff}} \tag{3.6.20}$$

This shows that the effective volume of the pcv with respect to the generation of marker shot noise is $V_{eff} = \pi a_i^2 L_i$, and the effective beam diameter is $2a_i$.

Examination of the arguments leading to (3.6.11) and (3.6.13) suggests the generalization

$$R_{msn}(\tau) = \frac{\iiint_{pcv} \varphi_i(x,y,z)\, \varphi_i(x+\tau\bar{U},y,z)\; dx\, dy\, dz}{\iiint_{pcv} \varphi_i^2\,(x,y,z)\; dx\, dy\, dz} \tag{3.6.21}$$

where x is the direction of motion and φ_i is the energy distribution function of the incident beam. This expression can be used to predict the correlation and spectral density function for cases of a laser-type of incident beam:

1. The velocity vector $\bar{\mathbf{U}}$ is parallel to the incident beam. The forms of $R_{msn}(\tau)$ and $H_{msn}(\tau)$ are the same as in the case of a uniform incident beam.
2. The pcv is right-cylindrical and its axis is normal to $\bar{\mathbf{U}}$. Then (3.3.7) gives

$$\varphi_i(x,y,z) = \frac{2}{\pi a_i^2}\, e^{-2(x^2+y^2)/a_i^2} \tag{3.6.22}$$

and (3.6.21) reduces to

$$R_{msn}(\tau) = \frac{4\int_0^\infty \int_0^\infty \varphi_i(x,y)\varphi_i(x+\tau\bar{U}, y)\,dx\,dy}{4\int_0^\infty \int_0^\infty \varphi_i^2(x,y)\,dx\,dy} = e^{-(\tau\bar{U}/a_i)^2} . \quad (3.6.23)$$

The corresponding spectral density function is

$$H_{msn}(f) = 2\sqrt{\pi}\,\frac{\overline{i^2_{msn}}\,a_i}{\bar{U}}\; e^{-(\pi f a_i/\bar{U})^2} \quad (3.6.24)$$

This result also applies to the case of an oblique-cylindrical pcv with its ends parallel to, and its axis normal to, $\bar{\mathbf{U}}$.

It is interesting to compare the limiting forms of the spectral density function as $f \to 0$. Equation (3.6.12) gives

$$H_{msn}(f) \approx \frac{2\overline{i^2_{msn}}L_i}{\bar{U}}\left[1 - 2\left(\frac{\pi f L_i}{\bar{U}}\right)^2\right] , \quad (3.6.25)$$

$2\pi f L_i/\bar{U} << 0.3$ roughly, and (3.6.24) gives

$$H_{msn}(f) \approx 2\sqrt{\pi}\,\frac{\overline{i^2_{msn}}\,a_i}{\bar{U}}\left[1 - \left(\frac{\pi f a_i}{U}\right)^2\right] , \quad (3.6.26)$$

$\pi f a_i/\bar{U} < 0.3$. This suggests that generally

$$H_{msn}(0) \approx 2\overline{i^2_{msn}}D_V/\bar{U} . \quad (3.6.27)$$

The effect of a particle size distribution on $\overline{i^2_{msn}}$ is evident from (3.6.16):

$$\overline{i^2_{msn}} = V\bar{\Gamma}^*\int_0^\infty s_p^{*2} m_p g(D_p)\,dD_p , \quad (3.6.28)$$

where s_p^* is the current sensitivity to particles of diameter D_p, and $g(D_p)dD_p$ is the mass fraction of particles in the diameter range D_p, $D_p + dD_p$. The definition of s_p^* is given by

$$\bar{I}^* = s^*V\bar{\Gamma}^* = V\bar{\Gamma}^*\int_0^\infty s_p^*\, g(D_p)\,dD_p . \quad (3.6.29)$$

For particles with $D_p > 0.2$ and light sources in the visible range, $\pi D_p/\lambda > 1$; the scattering efficiency of such particles is then expected to be of the order of magnitude of unity. Thus $s_p^* \propto A_p/m_p$, roughly, and in (3.6.28)

$$s_p^{*2} m_p \propto A_p^2/m_p \propto D_p .$$

This indicates that a given mass of large particles contributes more heavily to $\overline{i^2_{msn}}$ than an equal mass of small particles, the relative contributions being roughly as the ratio of the particle diameters. Hence, if one writes, from (3.6.2),

$$\overline{i^2_{msn}} = \bar{I}^{*2}/\bar{N}^*_{eff} \tag{3.6.30}$$

$\bar{N}^*_{eff}$ may be found to be considerably smaller than the actual number of particles in the pcv. Equation (3.6.2) also gives an expression defining an effective particle diameter:

$$\overline{i^2_{msn}} = \frac{\pi}{6}\rho_p D^3_{p,eff}\bar{I}^{*2}/V\bar{\Gamma}^* \ . \tag{3.6.31}$$

Shaughnessy [24] obtained (3.6.24) by applying Campbell's theorem [32] to evaluate the autocorrelation function,

$$R_{msn}(\tau) = A\int_{-\infty}^{\infty} \omega(t)\,\omega(t+\tau)\,dt \ , \tag{3.6.32}$$

where A is a constant for a given system, and $\omega(t)$ is the response of the phototransducer to a single particle passing through the pcv. This approach is generally valid for the evaluation of $R_{msn}(\tau)$. Care must be exercised that due allowance is made for differences between various paths through the pcv.

Marker shot noise has been negligible in all the author's applications of marker nephelometry, in the sense that it has been small compared to electronic shot noise. It is not difficult, however, to bring it into prominence by using a powerful laser light source and sufficiently low particle concentrations in the pcv. Shaughnessy [24], using a 2 W argon ion laser, a pcv diameter D_V of about 0.1 mm, and various concentrations of dioctyl phthalate condensation smoke particles was able to substantiate (3.6.20) experimentally. He also obtained data demonstrating the spectrum function (3.6.24).

The following conclusions are indicated by the preceding paragraphs:

1. Marker shot noise becomes significant when the incident beam is powerful, but the number of marker particles in the pcv is small. It then exceeds the electronic shot noise, and tends to be the limiting random noise factor in the detection of concentration fluctuations, particularly in respect to the spectral density function at high frequencies.

2. The characteristics of marker shot noise depend upon the particle properties, the geometry of the pcv, and the direction of the mean fluid velocity relative to the pcv. Marker shot noise can be calibrated for by studying the random noise in the nephelometer signal in a situation where the region being probed is steady and uniform in velocity and marker concentration. Either the marker concentration or the pcv size – or both – must be varied, and in the latter case the pcv's must be geometrically similar. The pcv geometry, it should be noted, involves the incident beam energy distribution $\varphi(r)$, the incident optical depth L_i, and the scattering angle θ_s. The results of such a calibration are (i) an expression for $\overline{i^2_{msn}}$, obtained by combining (3.4.7) and (3.6.31), of the form

$$\overline{i^2_{msn}} = s^* m_{p,eff}\bar{I}^*/V \equiv C_1\bar{I}^* \tag{3.6.33}$$

and (ii) an expression for $H_{msn}(f)$ of the form

$$H_{msn}(f) = C_2 \overline{i^2_{msn}} L h(fL/\bar{U})/\bar{U}$$

$$\equiv C_3 \overline{i^2_{msn}} h'(f/\bar{U})/\bar{U} \qquad (3.6.34)$$

where

$$A = \bar{U} H_{msn}(0)/\overline{i^2_{msn}} L \qquad (3.6.35)$$

is a constant and L is a suitable characteristic dimension of the pcv. The determination of C_1 and C_3 is straightforward. The further interpretation of these expressions requires that V and L be defined and that s^* be measured. The definition of V and L in respect to marker shot noise has been discussed earlier in this Section. The current sensitivity s^* is defined by (3.4.7) as $s^* \equiv \bar{I}^*/\bar{\Gamma}^*$, and an absolute measurement of $\bar{\Gamma}^*$ is therefore required. This can be done gravimetrically with the help of a sampling probe.

3. Marker shot noise can be used to determine an effective marker particle diameter $D_{p,eff}$ which is weighted towards the larger particles in a size distribution. This is an excellent and convenient way of determining whether the particle diameter is sufficiently small so that the marker will faithfully follow the motion of the marked stream (see Secs 3.11–3.16 for a discussion of marker fidelity).

3.7 *Electronic Shot Noise*

The theory of electronic shot noise is given in textbooks on electronics and certain special features associated with the more complicated types of phototransducers, particularly photomultiplier tubes, are discussed in manufacturers' literature [33,34]. The theory in connection with marker nephelometry has been examined by Becker *et al.* [20]. The general theory predicts that for a photodiode, or any single-stage device, the spectral density function of the electronic shot noise is given by

$$H_{esn}(f) = 2\epsilon\bar{I}\ , \qquad (3.7.1)$$

where ϵ is the charge on the electron and I is the total output current as given by (3.4.3).

In the case of a photomultiplier tube, most of the electronic shot noise arises at the photocathode – the first stage, where the current is smallest. The multiplier stages act mainly to amplify the cathode output, but a little additional noise is also introduced in these. The overall result is [33]:

$$H_{esn}(f) \approx \frac{2\epsilon}{\bar{I}_k} \frac{A_n}{A_n - 1}\ , \qquad (3.7.2)$$

where A_n is the amplification factor per stage. Usually $A_n = 5$, giving

$$H_{esn} \approx 2.5\,\epsilon\bar{I}^2/\bar{I}_k = 2.5\,A\epsilon\bar{I}\ , \qquad (3.7.3)$$

where A is the overall gain, approximately $A = A_n^N$, and N is the number of multiplier stages.

Electronic shot noise is kept small by working near the upper end of the dynamic range of the phototransducer, so that the mean current $\bar{I}$ is as large as possible. In any case, it is easy to calibrate a phototransducer for this noise. The calibration should be done in a dark room, or in any system such that the transducer receives light only from a very steady light source – a tungsten lamp powered by a battery is suitable. The electronic instrumentation to be used with the marker nephelometer must be part of the measurement system insofar as it determines the frequency pass-band. This applies to low-pass filters, spectral analyzers, and similar devices, or any set of instruments with its input and output cables and its given overall frequency response characteristics.

3.8 *Source Fluctuation Noise*

The radiant flux Φ_E entering the aperture of the exciter optics can be written as

$$\Phi_E = \bar{\Phi}_E(1+\beta) ,$$

where $\beta\bar{\Phi}_E$ represents the fluctuation about the mean due to random variation and ripple. The spectral density function of β is $B(f)$,

$$\overline{\beta^2} = \int_0^\infty B(f)\, df .$$

Then in (3.4.11) and (3.4.15)

$$\overline{i_E^2}\Big|_{f,\Delta f} = \overline{\beta^2}\Big|_{f,\Delta f} \bar{I}_s^{*2} \tag{3.8.1}$$

$$H_E(f) = B(f)\, \bar{I}_s^{*2} . \tag{3.8.2}$$

3.9 *Other Sources of Electrical Noise*

It is beyond the present scope to consider noise introduced by instruments used with the marker nephelometer probe – amplifiers, voltmeters, spectrum analyzers, etc.. One noise source that is common to all systems may be mentioned, however. The output current I from the phototransducer is generally passed through a signal resistor R, and the instruments respond to the voltage signal $\Delta E = IR$. This resistor introduces the fluctuation noise, of thermal origin, called *Johnson noise.* The maximum Johnson noise output from a resistor in a frequency interval, f, $f + \Delta f$ is equivalent to a mean-square current fluctuation

$$\overline{i_T^2}\Big|_{f,\Delta f} = 4kT\Delta f/R , \tag{3.9.1}$$

where k is the Boltzmann constant. This noise is added to the mean-square current fluctuation given by (3.4.10). The spectral density function of i_T is

$$H_T(f) = 4kT/R \ . \tag{3.9.2}$$

Johnson noise is normally inconsequential.

3.10 *Marker Adequacy: General Considerations*

It is normally required that the marker particles should follow the motion of the marked fluid as faithfully as possible. The marker properties should be virtually invariant over the field under study; the effects of coagulation, evaporation or sublimation, chemical reaction, and changes in index of refraction must therefore be negligible. The particles should everywhere constitute a negligible mass fraction of the material, and they should not significantly affect fluid properties such as the effective viscosity. Theoretical considerations relating to some of these requirements have been given by Rosenweig *et al.* [19] and Becker *et al.* [20], and are here discussed in Secs 3.11–3.16. The general theory of particle-fluid and particle-particle interactions in turbulent flows has been developed by a number of authors [35-37].

3.11 *Marker Inertia*

One requirement that is sometimes difficult to meet when all the others have been satisfied is that of small particle inertia; this determines the ability of the marker particles to follow the motion of the fluid. Rosenweig *et al.* [19] and Becker *et al.* [20] approached this problem by considering the case of a spherical particle in a sinusoidal fluid velocity field

$$u = \sqrt{2}\,\hat{u} \sin 2\pi f t \ , \tag{3.11.1}$$

where $\hat{u} \equiv \sqrt{\overline{u^2}}$.† Suppose that the fluid is a gas, the particle is acted upon by Stokes drag with the Cunningham correction, and weight force is negligible relative to drag. The equation of motion is then

$$m_p \frac{dv}{dt} = 3\pi\mu D_p (v - u)/(1 + K_g \ell/D_p) \ , \tag{3.11.2}$$

where $m_p = (\pi/6)\rho_p D_p^3$ is the particle mass, K_g is the Cunningham correction factor, 1.8 for air, ℓ is the molecular mean free path, and ℓ/D_p is the Knudsen number. The solution of this equation gives for the velocity amplitude ratio

$$\hat{v}/\hat{u} = [1 + (2\pi f/a)^2]^{-1/2} \tag{3.11.3}$$

where $\hat{v} \equiv \sqrt{\overline{v^2}}$ and

$$a \equiv 18\mu/\rho_p D_p^2 (1 + K_g \ell/D_p) \ , \tag{3.11.4}$$

Becker *et al.* give the example of particles of density 1 g/cm³ in air at 25°C and 1 atm, and find the following values of the frequency f at which $\hat{v} = 0.9\,\hat{u}$:

---

† The statement of (3.11.1) in [20] has a misprint, μ appearing in the place of π.

D_p, μm	0.1	0.3	1	3	10
f, khz	930	180	22	2.7	0.25

Doubling the viscosity doubles this frequency, and doubling the particle density halves it.

Estimates for liquids can be made from (3.11.1) by substituting

$$a \equiv 18\mu/D_p^2(\rho_p + \tfrac{1}{2}\rho) \tag{3.11.5}$$

where ρ is the fluid density, and the addition of $\tfrac{1}{2}\rho$ to ρ_p accounts for the virtual inertia of the particle (the concept of virtual inertia, or virtual mass, is discussed by Birkhoff [38] and Batchelor [39]).

3.12 *Gravitational Drift*

The effect of particle weight will be small if the terminal fall velocity is small compared with velocities in the field whether fluctuating or mean. The terminal speed v_t of a particle acted on by Stokes drag with the Cunningham correction is

$$v_t = gD_p^2(\rho_p - \rho)(1 + K_g\ell/D_p)/18\mu \tag{3.12.1}$$

where ρ is the fluid density. For the example considered following (3.11.1), v_t equals 0.35 mm/s. A rigorous criterion for the neglect of this effect is $v_t t < \lambda_u/3$, where t is the transit time of a particle from its entrance into the field up to a given point, λ_u is the Kolmogoroff microscale $\lambda_u \equiv (\nu^3/\epsilon_T)^{1/4}$ at that point, $\nu \equiv \mu/\rho$ is the kinematic viscosity, and ϵ_T is the rate of viscous dissipation of turbulence kinetic energy.

3.13 *Marker Coagulation*

The coagulation of sol particles in a fluid at rest often follows a second-order rate expression [40]:

$$-\frac{d\mathcal{C}}{dt} = k_r\mathcal{C}^2 , \tag{3.13.1}$$

where $\mathcal{C}$ is the number concentration of particles. The rate constant k_r for sticky spherical particles is given by Smoluchowski's theory for gases as

$$k_r = \frac{4}{3}kT(1 + K_g\ell/D_p)/\mu . \tag{3.13.2}$$

Consider, then, a uniformly marked fluid in which the marker particles are initially monodisperse spheres. The above expressions will be applied in the neighborhood of $t = 0$ (where the majority of collisions are between the original particles) to obtain the initial rate of coagulation. We have $\mathcal{C}^* = \Gamma^*/m_p = 6\Gamma^*/\pi\rho_p D_p^3$, and (3.13.1) thus gives

$$\frac{dD_p}{dt} = \frac{2}{\pi}\frac{k_r\Gamma^*}{\rho_p D_p^2} \tag{3.13.3}$$

From (3.4.6), the response of a marker nephelometer to this system is $I^* = s^*\Gamma^*$. If the particles satisfy $\pi D_p/\lambda > 1$, then, as suggested following (3.6.29), s^* is roughly proportional to D_p^{-1}, giving

$$\frac{ds^*}{dt} = -\frac{s^*}{D_p}\frac{dD_p}{dt} \tag{3.13.4}$$

for small variations in D_p and s^*. Combination of this expression and (3.13.3) gives

$$\frac{d \ln s^*}{dt} \approx -\frac{k_r\Gamma^*}{3m_p} = -\frac{d \ln D_p}{dt} . \tag{3.13.5}$$

If, on the other hand, the particles are very small dielectric spheres giving Rayleigh scatter $(\pi D_p/\lambda << 1)$ then $s^* \propto D_p^3$ and

$$\frac{d \ln s^*}{dt} \approx \frac{k_r\Gamma^*}{m_p} = 3\frac{d \ln D_p}{dt} \tag{3.13.6}$$

It is expected that (3.13.5) should normally apply in marker nephelometry, $\pi D_p/\lambda$ being too large for (3.13.6). It should be noted that these expressions give only an order-of-magnitude estimate of the maximum coagulation rates.

The effect of turbulence is to increase the rate of coagulation. The equation of Saffman and Turner [35] for the coagulation rate due to the action of turbulence alone in an isotropic turbulent flow gives

$$-\frac{d\mathcal{C}}{dt} = \frac{2}{3} D_p^3 \mathcal{C}^2 \left\{\frac{2\pi\epsilon_T}{\nu}\right\}^{1/2} \tag{3.13.7}$$

Then

$$\frac{dD_p}{dt} = 1.06 \frac{D_p\Gamma^*}{\rho_p}\left\{\frac{\epsilon_T}{\nu}\right\}^{1/2} \tag{3.13.8}$$

and if $\pi D_p/\lambda > 1$,

$$\frac{d \ln s^*}{dt} \approx -1.06 \frac{\Gamma^*}{\rho_p}\left\{\frac{\epsilon_T}{\nu}\right\}^{1/2} = -\frac{d \ln D_p}{dt} . \tag{3.13.9}$$

Equations (3.13.3) and (3.13.8) can be added to estimate the combined effects of Brownian motion and turbulence.

The value of ϵ_T can be estimated from Kolmogoroff's second hypothesis in the form used by Spalding and coworkers [41-43]:

$$\epsilon_T \approx 0.1\, e_K^{3/2}/\Lambda_u \tag{3.13.10}$$

where $e_K \equiv \frac{1}{2}(\overline{u_x^2} + \overline{u_y^2} + \overline{u_z^2})$ is the turbulence kinetic energy and Λ_u is the integral length scale of the stream-wise velocity fluctuations.

All the above results have been developed as though the fluid were in a cell, either stagnant or stirred by isotropic turbulence, and uniform in Γ^*. In applying the equations to fields where flow and mixing are occurring, the time derivatives are to be treated as Stokes derivatives and the equations are to be suitably averaged. Equation (3.13.9), for example, can be written

$$\frac{D \ln s^*}{Dt} = -1.06 \frac{\bar{\Gamma}^*}{\rho_p} \left\{ \frac{\epsilon_T}{\nu} \right\}^{1/2} = -\frac{D \ln D_p}{Dt} . \qquad (3.13.11)$$

Consider, as an example, a point 50 cm downstream in a free jet of air at 25°C and 1 atm from a 1 cm dia. nozzle. Suppose that $\hat{u}_x = 3$ m/s, $\bar{\mathfrak{C}}^* = 10^6$ particles/cm^3, $D_p = 1$ μm, and $\rho_p = 1$ g/cm^3. Roughly, $e_K = 13$ m^2/s^2, and [21], $\Lambda_u = 0.07x = 3.5$ cm. We have $\bar{\Gamma}^* = m_p \bar{\mathfrak{C}}^* = 0.005$ kg/m^3. Then (3.13.5) gives

$$D(\ln D_p)/Dt = -D(\ln s^*)/Dt = 0.00011 \text{ s}^{-1}$$

for coagulation by Brownian motion, and (3.13.9) gives

$$D(\ln D_p)/Dt = -D(\ln s^*)/Dt = 0.016 \text{ s}^{-1}$$

for coagulation by turbulent motion. These coagulation rates appear to be reasonably low, and should cause no difficulty. To solve the problem most rigorously, though, the equations should be integrated from the point of entry to the field – the nozzle, along the mean-flow pathline up to the point in question Suppose that point is on the jet centreline; then (3.13.11) gives

$$\ln(D_p/D_{p,o}) = \frac{1.06}{\rho_p} \int_0^t \bar{\Gamma}^* \left\{ \frac{\epsilon_T}{\nu} \right\}^{1/2} dt = \frac{1.06}{\rho_p} \int_0^x \frac{\bar{\Gamma}^*}{\bar{U}_x} \left\{ \frac{\epsilon_T}{\nu} \right\}^{1/2} dx \qquad (3.13.12)$$

where $\bar{\Gamma}^*$, $\bar{U}_x$ and ϵ_T are evaluated along the jet centreline.

3.14 Vaporization or Dissolution of Particles

The rate of disappearance of a marker particle by vaporization in a gaseous medium or dissolution in a liquid is

$$-dm_p/dt = k_z S_p (z - z_\infty) \qquad (3.14.1)$$

where z is the fugacity or activity of the marker material at the interface in the medium (or a measure thereof in terms of pressure, concentration, or composition) and k_z is the mass-transfer coefficient. The Reynolds number Re of a particle relative to the surrounding medium is generally very small, and so the Sherwood number Sh (a dimensionless measure of K_z) takes the constant value that is characteristic of the limit at low Re where the mass transfer rate is determined purely by diffusion.

In the case of particles in a gaseous medium it is usually convenient to write (3.14.1) as

$$-dm_p/dt = k_p M_v S_p (p_v - p_\infty) , \tag{3.14.2}$$

where p_v is the vapour pressure of the particle material at the particle surface, p_∞ is the vapour pressure in the surrounding gas, and k_p is the transfer coefficient based on the vapour pressure driving force. The corresponding Sherwood number for spherical particles at low Re has the value

$$\mathrm{Sh} \equiv \frac{k_p P D_p}{c D_v} = \frac{k_p R T D_p}{D_v} = 2 , \tag{3.14.3}$$

where P is the total gas pressure, c is the gas molar density, R is the gas constant, and D_v is the vapour diffusivity in the gas. Thus $k_p = 2D_v/RTD_p$ and (3.14.2) gives

$$-\frac{d \ln D_p}{dt} = \frac{4M_v D_v p}{RT\rho_p D_p^2} \frac{p_v - p_\infty}{P} . \tag{3.14.4}$$

In order to estimate the vapour pressure p_v in (3.14.4), it is necessary that the particle temperature T_p be known. The Biot number Bi should generally be very small, because the particles are so small, and hence the heat transfer rate is gas-phase controlled. A quasi-steady-state energy balance is sufficient to estimate the maximum temperature difference between a particle and the ambient gas:

$$hS_p(T_\infty - T_p) = k_p M_v S_p \Delta h_v (p_v - p_\infty) , \tag{3.14.5}$$

where h is the heat transfer coefficient, and Δh_v is the specific enthalpy of vaporization. In the limit at low Reynolds numbers $h = 2k/D_p$, where k is the thermal conductivity of the gas. Thus (3.14.5) gives, with (3.14.3),

$$T_p = T_\infty - \frac{D_v M_v \Delta h_v}{kRT}(p_v - p_\infty) = T_\infty \frac{M_v \Delta h_v}{MC_p \,\mathrm{Le}} \frac{p_v - p_\infty}{P} \tag{3.14.6}$$

where $\mathrm{Le} \equiv a/D_v \equiv k/\rho C_p D_v$ is the Lewis number, M is the molecular weight of the gas, and C_p is the specific heat capacity of the gas. Since p_v is itself a function of T_p, this is an implicit form of the solution.

In the case of liquid droplets, the vapour pressure p_v is affected by the surface curvature:

$$p_v = p^o \exp(4\,\sigma M_v/RT\rho_p D_p) , \tag{3.14.7}$$

where p^o is the normal vapour pressure, and σ is the surface tension. The enthalpy of vaporization of organic liquids is roughly given by Trouton's rule, $M_v \Delta h_v = C_T T_b$, where T_b is the boiling point at 1 atm in kelvins, and $C_T = 89$ J/g-mole.

The author has used oil condensation smoke markers in air at 25°C and 1 atm in most of his work. The oils used [20] have a mean molecular weight M_v in the range 250-300 g/g-mol, a normal vapour pressure

$$p^o \;=\; 1.17 \times 10^5 \exp(-8220/T) \text{ atm} , \tag{3.14.8}$$

a vapour diffusivity in air at 25°C and 1 atm of $\mathcal{D}_v = 0.03$ cm²/s, a liquid density $\rho_p = 900$ kg/m³, and a surface tension $\sigma \approx 37$ mN/m. Consider a droplet of 1 μm dia. Equations (3.14.6), (3.14.7) and (3.14.8) yield $p^o = 1.23 \times 10^{-7}$ atm, $p_v = 1.02\, p^o = 1.25 \times 10^{-7}$ atm, and negligible difference between T_p and T_∞. Then (3.14.4) gives for $p_\infty = 0$, $d(\ln D_p)/dt = 0.02\ \text{s}^{-1}$. This rate of diameter change should not be serious so long as the transit times across the field under study are well under 1 s.

3.15 *Thermal Effects of the Incident Beam*

It is often said that optical probes do not interfere with a flow. This may not be true, however, when very powerful laser beams are used. Under intense irradiation, marker particles in a gaseous medium may be appreciably heated, changing their index of refraction, increasing their rate of vaporization in the case of liquid droplets, etc.. This heating effect can be estimated from an energy balance on a particle. The differential equation for transient heating is

$$m_p C_p \frac{dT_p}{dt} \;=\; \eta_a A_p G_i - S_p h (T_p - T) , \tag{3.15.1}$$

where η_a is the absorption efficiency,

$$\eta_a G_i \;=\; \int_0^\infty \eta_{\lambda a} G_{\lambda i}\, d\lambda , \tag{3.15.2}$$

S_p, C_p and T_p are respectively the particle's surface area, heat capacity, and temperature, T is the gas temperature, h is the heat transfer coefficient, and G_i is the incident radiant flux density as defined in Sec.3.3. If the particles are small, as is usual, and if T_p and T are such that radiant heat transfer is unimportant, then $h \approx 2k/D_p$, where k is the thermal conductivity of the fluid. The effect of the Knudsen number ℓ/D_p is here neglected. The solution of (3.15.1) for a spherical particle is

$$T_p - T \;=\; \frac{\eta_a D_p G_i}{8k}\,(1 - e^{-\beta t}) \tag{3.15.3}$$

$$\beta \;\equiv\; 12\, k/\rho_p D_p^2 C_p \,. \tag{3.15.4}$$

As an example, consider oil droplets with $D_p = 1$ μm, $\rho_p = 0.9$ g/cm³, $C_p = 2$ J/kg K, in air at 25°C, 1 atm, under a laser beam of 1 W power and 1 mm effective diameter. It is calcualted that

$$G_i \;=\; 1.27 \times 10^6 \text{ W/m}^2$$

$$D_p G_i/8k = 6.1°\text{C}$$

$$\beta = 173000 \text{ s}^{-1} \quad .$$

The maximum value of the absorption efficiency η_a is of the order of unity. If the air flow is at right-angles to the incident beam and the air velocity is 1 m/s, the maximum residence time of a particle in the beam is 1 ms. Then $\beta t = 173$, and for $\eta_a = 1$, $(T_p - T) = 6.1°$C. When the same laser beam is focussed to 0.1 mm dia., the answers are $D_p G_i/8k = 610°$C, $\beta t = 17.3$, and for $\eta_a = 1$, $(T_p - T) = 610°$C. The high values of βt indicate that particles can usually be expected to reach the steady-state temperature given by (3.15.3) within the pcv, since $\exp(-\beta t) \ll 1$. These estimates of $(T_p - T)$ may, of course, be high, because η_a may be significantly smaller than unity. Nevertheless, it is evident that if $G_i < 10^6$ W/m^2, the effect should be negligible, but if $G_i > 10^7$ W/m^2, it may be quite appreciable.

3.16 Photophoresis

Another probe interference effect that may be important in gases is the particle drift caused by the absorption of photon momentum, an effect called *photophoresis.* The effect can be estimated from a force balance on a spherical particle starting from rest in a still gas and effectively acted upon only by the net radiation force due to the incident beam and by Stokes drag with the Cunningham correction. The energy flux intercepted by the particle is $\eta A_p G_i$,

$$\eta G_i \equiv (\eta_s + \eta_a) G_i \equiv \int_0^\infty (\eta_{\lambda s} + \eta_{\lambda a}) G_{\lambda i} \, d\lambda \ , \tag{3.16.1}$$

where η, η_s and η_a are respectively the extinction efficiency, the scatter efficiency, and the absorption efficiency. The intercepted momentum flux is therefore $\eta G_i/c$, where c is the speed of light. The scattered energy flux is $\eta_s A_p G_i$, and its component of momentum flux in the direction of the incident beam is $\eta_s p_i A_p G_i/c$, where

$$p_i = \int_{2\pi} p \cos\theta \, d\Omega = \int_{2\pi} \int_0^\infty p_\lambda \cos\theta \, d\lambda \, d\Omega_s \ . \tag{3.16.2}$$

The resulting net radiation force acts in the direction of the incident beam, and its magnitude is

$$\eta_m A_p G_i/c \equiv (\eta - \eta_s p_i) A_p G_i/c \ . \tag{3.16.3}$$

The equation of motion of the particle can thus be written as

$$m_p \frac{dv}{dt} = \eta_m A_p G_i/c - 3\pi\mu v D_p / (1 + K_g \ell/D_p) \ . \tag{3.16.4}$$

The solution of this equation is

$$v = \frac{\eta_a D_p G_i(1 + K_g \ell/D_p)}{12\mu c} (1 - e^{-at}) \,, \qquad (3.16.5)$$

giving

$$s = \int_0^t v\,dt = vt\left\{1 - \frac{1}{at}(1 - e^{-at})\right\} \,, \qquad (3.16.6)$$

where a is given by (3.11.4).

For the example considered following (3.11.3) (namely particles of density $1\ g/cm^3$ and diameter $1\ \mu m$ in air at 25°C and 1 atm) and for a 1 mm dia., 1 W laser beam it is found that $a = 2.9 \times 10^5\ s^{-1}$ and

$$D_p G_i(1 + K_g \ell/D_p)/12\mu c = 2.2\ mm/s \,.$$

For radiation in the visible range, $\pi D_p/\lambda \approx 5$ for the given particles. This is a moderately large value of $\pi D_p/\lambda$, and one can therefore take $\eta_m = 1$ as an order-of-magnitude estimate. The calculated results for an exposure time t of 1 ms are then $at = 290$, $v = 2.2$ mm/s, and $s = 2.2\ \mu m$. The very large value of at indicates that particles can usually be expected to approach the equilibrium photophoretic drift velocity within the pcv. The effect estimated in this example is negligible. When the same laser beam is focussed to 0.1 mm dia., however, it is estimated that $v = 22$ cm/s, and $s = 0.22$ mm, which might sometimes be significant.

It appears from these calculations that photophoretic drift should usually be negligible if $G_i < 10^6\ W/m^2$, but difficulties may well be encountered at $G_i > 10^8\ W/m^2$.

3.17 *Effects of Optical Extinction and Background Radiation*

The nephelometer response equations, e.g. (3.4.6) and (3.4.11), show that the response is modulated by optical extinction along the paths of the incident and scattered beams. The effective transmissivity τ for the whole path can generally be written as $\tau = \tau^\circ \tau'$, where τ° is the transmisivity of the medium (gas or liquid) under control conditions, and τ' is the extra effect under the actual experimental conditions. Normally τ° is virtually unity in gaseous systems, but may be significantly less in liquid systems. It is a constant for a given path, and it should usually be easy to calibrate the system to define τ° for each path in question.

The value of the perturbation extinction factor $1 - \tau'$ should be held as small as possible, desirably so small that corrections for its effect are unnecessary. The dividing line obviously depends upon the desired accuracy, but usually "good" operating conditions require that $1 - \overline{\tau'} < 0.02$, and "fair" conditions that $1 - \overline{\tau'} < 0.05$.

The effects of variable extinction and of background radiation are difficult to discuss in a general way, as has already been indicated in connection with (3.4.13). Various critical experiments can usually be devised to assess these effects to some degree, e.g., direct measurement of extinction along certain paths, variation of the input marker concentration, comparison of measurements at the same point using different optical paths, and checks on symmetry of results for fields that have elements of symmetry. Any effects for which quantitative corrections cannot be developed should be minimized as far as possible consistent with other constraints, e.g., marker shot noise.

When marker concentration fluctuations are of interest, as well as the mean concentration field, then the unknown degree of correlation between these fluctuations, fluctuations in τ', and fluctuations in certain types of background radiation (e.g. in-scatter into the aperture of the detector) can be a major source of uncertainty. When all other approaches to reducing uncertainty to an acceptable level are deemed inadequate, the last resort is to shield the incident and scattered beams physically and to block the detector's view of the background beyond the pcv. Raichura [44] and Rajani [45], studying jets of dust, used glass tubes to conduct the incident and scattered beams. Yang and Meroney [46, 47] and Liu and Karaki [48] have designed probes with fibre optic conduits. The author has made use of a water-cooled fibre-optic system in measurements on a flame.

4. INTERPRETATION OF MARKER CONCENTRATION FIELDS

4.1 General

Marker nephelometry can obviously be used for the direct study of sol concentration fields, e.g., in experiments on air pollution. If the marker particles have the same diffusional characteristics in the given fluid as the particles of interest – or if those very particles constitute the marker – then an accurate representation of the transport of sol particles will be obtained.

More frequently, however, it is desired to use a marker to follow molecular processes of mixing and dilation within fluids. Suppose that one of the streams entering a given field is uniformly marked with sol particles. The mass concentration of material from the marked feedstream at any point will be denoted by Γ. By "material of the marked stream" is meant, in general, the atomic material of that stream, which is conserved; the molecular matter, by contrast, may be transformed by chemical reactions in the field. If the binary diffusivities of all pairs of molecules in the field were equal, and if the diffusivity of the marker in the field were the same as these molecular diffusivities, then the marker would exactly portray the transport of material of the marked stream:

$$\Gamma/\Gamma_o = \Gamma^*/\Gamma_o^* , \tag{4.1.1}$$

where Γ_o $(=\rho_o)$ and Γ_o^* are the input mass concentrations of marked fluid and marker. The marker would also depict the transport of the material of any component or complex of components introduced with the marked stream,

$$\Gamma_C/\Gamma_{C,o} = \Gamma^*/\Gamma_o^* . \tag{4.1.2}$$

In a combustion system, for example, "*C*" might denote the fuel atoms or the carbon atoms in a marked gaseous fuel feedstream, or oxygen atoms in an airstream.

Consideration hereafter, unless otherwise noted, will be restricted to gaseous systems, mainly because most of the work on marker nephelometry has been done with gases [22,24,44-48], generally with air at room temperature. The modifications required for liquid systems will be fairly obvious, and need no special comment.

Equality of binary molecular diffusitivites in gases is usually approximated to a fair degree, the Schmidt number $\mathrm{Sc} \equiv \nu/\mathcal{D}$ being about 0.7 for diatomic gases; only for a few exceptional species such as the H radical is Sc very markedly different from unity. When a turbulent system is to be treated, the problem is normally complicated enough, and enough approximations of a similar crudeness have to be made, so that the assumption of equal binary diffusivities is quite acceptable and perhaps the only practical approach. It is easily shown [49] from the Stefan-Maxwell diffusion equations that when the binary diffusivities $\mathcal{D}$ are equal, and the mixture molecular weight M is virtually a constant, then atomic material from a given feedstream is effectively transported as a complex† under the diffusivity $\mathcal{D}$:

$$\rho DW/Dt = \nabla\cdot\mathcal{D}\rho\nabla W , \tag{4.1.3}$$

where $W \equiv \Gamma/\rho$ is the mass fraction of material of that feedstream. An analogous equation applies to $W_C \equiv \Gamma_C/\rho$. It is required, strictly speaking, that the given feedstream be uniform in composition; however, this condition can be relaxed if the feedstream acts virtually as a point or line source of material with respect to the system.

It is evident, then, that a suitable gaseous marker introduced with one of the feedstreams entering a gaseous system will give a good representation of the transport of material of that stream. It must now be asked how good a representation is obtained when the marker consists of sol particles whose diffusivity $\mathcal{D}^*$ is considerably different from that of gaseous species. The diffusivity of sol particles in a gas is given by the Stokes-Einstein equation with Cunningham's correction [40],

$$\mathcal{D}^* = kT(1 + K_g\ell/D_p)/3\pi\mu D_p . \tag{4.1.4}$$

For particles of 1 μm dia. in air at 25°C and 1 atm, $\mathcal{D}^* = 2.7 \times 10^{-7}$ cm²/s and $\mathrm{Sc}^* = 570{,}000$. Thus $\mathrm{Sc}^* >\!\!>\!\!> 1$, rather than the desired condition $\mathrm{Sc}^* \approx 1$.

As far as the mean concentration fields Γ and Γ^* are concerned, it is well known that in a flow of high turbulence Reynolds number $\Lambda u/\nu$ the turbulent diffusivity $\mathcal{D}_T = \Lambda u$ in any case greatly exceeds the molecular diffusivity $\mathcal{D}$, where Λ is an integral length scale of the turbulence and u, the intensity of velocity fluctuation, can be taken to be $u = e^{1/2}$. Thus the fields of Γ and Γ^* in a well-developed turbulent flow are essentially determined by the turbulent transport, and the influence of $\mathcal{D}$ and $\mathcal{D}^*$ tends to be quite negligible. Both theory and experiment show that it can usually be assumed that

† Atoms of the given feedstream can effectively diffuse together with diffusivity $\mathcal{D}$ regardless of the molecules to which they may be tied by chemical reactions.

$$\Gamma/\Gamma_o \;\equiv\; \Gamma/\rho_o \;=\; \Gamma^*/\Gamma_o^* \tag{4.1.5}$$

A similar result can be written for an atomic species, or a complex of such species, that is present in the marked stream;

$$\Gamma_C/\Gamma_{C,o} \;\equiv\; \Gamma_C/W_{C,o}\rho_o \;=\; \Gamma^*/\Gamma_o^* \tag{4.1.6}$$

Equation (4.1.6) is essentially a corollary of (4.1.5) under the assumption of equal binary molecular diffusivities; the result for the complex will therefore not be written hereafter but will be taken as implied.

A somewhat different situation is found with respect to concentration fluctuations. Obviously diffusion must be important at sufficiently high wave numbers in the wave-number spectrum, and there the effect of the difference between $\mathcal{D}$ and $\mathcal{D}^*$ must be profound.

4.2 *Spectrum and Mean-square Value of Concentration Fluctuations*

The frequency and wave-number spectral density functions of the marker concentration fluctuations γ^*, namely $G^*(f)$ and $E^*(\kappa)$, were introduced in Sec.3.5. The corresponding functions for fluctuations in the concentration of material of the marked stream, γ, are $G(f)$ and $E(\kappa)$,

$$\overline{\gamma^2} \;=\; \int_0^\infty G(f)\,df \;=\; \int_0^\infty E(\kappa)\,d\kappa \;, \tag{4.2.1}$$

where $\kappa = 2\pi f/\bar{U}$, and $\bar{U} \equiv \sqrt{(\bar{U}_x^2 + \bar{U}_y^2 + \bar{U}_z^2)}$ is the magnitude of the mean velocity vector $\bar{\mathbf{U}}$. Consider the expected forms of E^* and E:

1. At low wave numbers, $\kappa << \Lambda_u^{-1}$, both E^* and E are expected to be independent of κ, where Λ_u is the integral length scale of the component of velocity fluctuation in the direction of $\bar{\mathbf{U}}$ – the "longitudinal" length scale.

2. In approximately the range $\kappa_m < \kappa < \kappa_u$, both E^* and E usually follow a power-law relation,

$$E^*(\kappa) \;=\; A^*\kappa^{-m} \tag{4.2.2}$$

$$E(\kappa) \;=\; A\kappa^{-m} \tag{4.2.3}$$

where $\kappa_m \approx 2\Lambda_u^{-1}$, $\kappa_u \equiv \lambda_u^{-1}$, and $\lambda_u \equiv (\nu^3/\epsilon_T)^{1/4}$ is the Kolmogoroff length scale. The value of the exponent m [21,51,52] is usually between 5/3 and 2. This type of power-law behaviour is associated with the convection subregime of the spectrum in which the energy cascade from low to high wavenumbers is inertially controlled, with $m = 5/3$ in the case of the Kolmogoroff equilibrium structure in which the turbulence is supposed to be essentially isotropic over the wavenumber range of the law.

3. The maximum in the dissipation spectrum of velocity fluctuations is around $\kappa = \kappa_u \equiv \lambda_u^{-1}$, whereas the maxima in the dissipation of concentration fluctuations, according to the theory of Batchelor [53], are around $\kappa = \kappa_\gamma \equiv \lambda_\gamma^{-1}$ for γ and $\kappa = \kappa_\gamma^* \equiv \lambda_\gamma^{*-1}$ for γ^*, where

$$\lambda_\gamma \equiv (\nu \mathcal{D}^2/\epsilon_T)^{1/4}, \qquad \lambda_\gamma^* \equiv (\nu \mathcal{D}^{*2}/\epsilon_T)^{1/4} .$$

Because Sc ≈ 1 for gases, $\lambda_\gamma \approx \lambda_u$. It is accordingly expected that $E(\kappa)$ will drop rapidly with κ beyond $\kappa = \kappa_u$.

4. Since $\mathcal{D}^* << \nu$, and thus $\lambda_\gamma^* << \lambda_u$, the convective straining of marker concentration "eddies" must continue far beyond $\kappa = \kappa_u$ before dissipation becomes pronounced. According to Batchelor's theory [53]

$$E^*(\kappa) = B^* \kappa^{-1} \tag{4.2.4}$$

in the approximate range $\kappa_u < \kappa < \kappa_\gamma^*$. Finally, at $\kappa > \kappa_\gamma^*$, $E^*(\kappa)$ drops rapidly with increasing κ.

The following conclusions for well-developed turbulent flows are suggested by these expectations:

1. To a good approximation, at $\kappa < 0.5\kappa_u$

$$E(\kappa) = (\bar{\Gamma}/\bar{\Gamma}^*)^2 \, E^*(\kappa) . \tag{4.2.5}$$

The upper bound on κ here is a reasonable guess.

2. To a good approximation,

$$\overline{\gamma^2} = (\bar{\Gamma}/\bar{\Gamma}^*)^2 \int_0^{\kappa_u} E^*(\kappa)\, d\kappa . \tag{4.2.6}$$

3. The value of $E(\kappa)$ in its dissipation regime, $\kappa \approx \kappa_\gamma \approx \kappa_u$ and beyond, cannot be determined by marker nephelometry.

4. We have exactly

$$\overline{\gamma^{*2}} = \int_0^{\infty} E^*(\kappa)\, d\kappa . \tag{4.2.7}$$

This approximates to

$$\overline{\gamma^{*2}} = \int_0^{\kappa_m} E^*(\kappa)\, d\kappa + \int_{\kappa_m}^{\kappa_u} A^* \kappa^{-m}\, d\kappa + \int_{\kappa_u}^{\kappa_\gamma} B^* \kappa^{-1}\, d\kappa , \tag{4.2.8}$$

where κ_m is the lower bound of the power law (4.2.2), say $\kappa_m = 2\Lambda_u^{-1}$.

The following forms provide further useful approximations to $\overline{\gamma^2}$, and have been extensively used in the author's work:

5. Extrapolation of the power law (4.2.2) gives

$$\overline{\gamma^2} \approx (\bar{\Gamma}/\bar{\Gamma}^*)^2 \left\{ \int_0^{\kappa_m} E^*(\kappa)\, d\kappa + \int_{\kappa_m}^{\infty} A^* k^{-m}\, d\kappa \right\} . \tag{4.2.9}$$

6. Suppose that measurements are made with a pcv such that $D_V \approx \lambda_u$ or somewhat smaller. Then

$$\overline{\gamma^2} \approx (\bar{\Gamma}/\bar{\Gamma}^*)^2 \int_0^\infty G_V^*(f)\, df \equiv (\bar{\Gamma}/\bar{\Gamma}^*)^2 \int_0^\infty Q_\kappa E^*(\kappa)\, d\kappa \equiv \overline{\breve{\gamma}^{*2}} \tag{4.2.10}$$

The practical implications of (4.2.5–10) are best seen through an example. Becker *et al.* [20,21] have reported extensive data on a round free jet of air at 25°C and 1 atm at a point $x = 32D$ on the axis, where x is distance from the nozzle plane and $D = 0.623$ cm is the nozzle diameter. The value of $\overline{\gamma^2}$ was estimated from the spectrum by means of (4.2.9); this value will be taken as unity, and estimates from the other formulae will be compared with it. The parameters required in these estimates are obtained as follows:

1. The input data are: the sol (oil condensation smoke) particle diameter [19,20], $D_p = 0.5$ μm; the mean gas velocity [20], $\bar{U}_x = 30.5$ m/s; the specific turbulence kinetic energy [50], $e_K = 0.078\, \bar{U}_x^2$; and the integral length scale for velocity fluctuations [21], $\Lambda_u = 1.30$ cm.

2. Equation (4.1.4) gives $\mathcal{D}^* = 6 \times 10^{-7}$ cm²/s; thus Sc* = 260,000. The specific rate of dissipation of turbulence kinetic energy, ϵ_T, is given by the Kolmogoroff formula (3.13.10). Calculation then yields $\lambda_u = (\nu^3/\epsilon_T)^{1/4} = 29.7$ μm, and $\lambda_\gamma^* = (\nu \mathcal{D}^{*2}/\epsilon_T)^{1/4} = 0.058$ μm. Thus $\kappa_u \equiv \lambda_u^{-1} = 336$ cm^{-1}, and $\kappa_\gamma^* \equiv \lambda_\gamma^{*-1} = 172000$ cm^{-1}. We also have $\kappa_m \approx 2\Lambda_u^{-1} = 1.54$ cm.

3. Becker *et al.* [21] give experimental data on the spectrum in the normalized form $\tilde{E} = \tilde{E}(\eta)$, where $\eta \equiv \Lambda_\gamma \kappa$, $\tilde{E}(\eta) \equiv E(\kappa)/\Lambda_\gamma$, Λ_γ is the integral length scale of concentration fluctuations in the direction of $\bar{\mathrm{U}}$ – another "longitudinal" length scale, estimated by

$$\Lambda_\gamma = \frac{\pi}{2} E(0) \,/\, \overline{\gamma^2} \,, \tag{4.2.11}$$

and $\overline{\gamma^2}$, as already noted, is estimated in terms of $E(\kappa)$ by (4.2.9). The definitive equation for Λ_γ is (4.4.2). The estimated value of Λ_γ at the considered point is 0.90 cm. At $\eta > 2$,

$$E(\eta) = 0.395\, \overline{\gamma^2}\, \eta^{-5/3} \,. \tag{4.2.12}$$

Since the differences between (4.2.5–10) only enter in the high wave-number regions, (4.2.12) provides all the information needed about the spectrum to carry out the desired comparisons.

The estimates of $\overline{\gamma^2}$ obtained from the various formulae are shown in Table I. The solutions of (4.2.10) are based on the supposition that (4.2.12) applies at all high wave numbers, but the answers would only be very slightly different if the −1 power law, (4.2.4), were introduced beyond $\kappa = \kappa_u$. In order to indicate the sensitivity to error in the limits of integration, (4.2.6) has also been solved with the upper limit $\kappa = 0.5\, \kappa_u$, and (4.2.8) has been solved with limits of $0.5\, \kappa_u$ and $0.5\, \kappa_\gamma^*$.

Equation (4.2.6) theoretically provides the best estimate of $\overline{\gamma^2}$. It is then evident that (4.2.6) gives an excellent approximation, and so does (4.2.10) when $D_V < 0.2\, \Lambda_\gamma$. Comparison with the results from (4.2.8) shows that the difference

Table I. Comparison of values of $\overline{\gamma^2}$ calculated from various formulae for a point on the axis of a free round jet of air at 25° C and 1 atm; $D = 0.623$ cm, $x = 19.3$ cm, $\overline{U} = 30.5$ m/s.

Equation	$\overline{\gamma^2}$ or $\overline{\gamma^{*2}}$	Remarks
(4.2.9)	1.000	$\overline{\gamma^2}$, reference case
(4.2.6)	0.987	$\overline{\gamma^2}$
(4.2.6)	0.979	$\overline{\gamma^2}$, limit $0.5\ \kappa_u$ instead of κ_u
(4.2.10)	0.991	$\overline{\gamma^2}$, $D_V = 0.03\ \mu m = \lambda_u$
(4.2.10)	0.968	$\overline{\gamma^2}$, $D_V = 0.2$ mm
(4.2.10)	0.941	$\overline{\gamma^2}$, $D_V = 0.5$ mm
(4.2.10)	0.909	$\overline{\gamma^2}$, $D_V = 1$ mm
(4.2.8)	1.041	$\overline{\gamma^{*2}}$
(4.2.8)	1.066	$\overline{\gamma^{*2}}$, limits $0.5\ \kappa_u$ and $0.5\ \kappa_\gamma^*$ instead of κ_u and κ_γ^*

between $\overline{\gamma^2}$ and $\overline{\gamma^{*2}}$ is small.

Some general conclusions are fairly apparent from these results:

1. The mean-square gas concentration fluctuation $\overline{\gamma^2}$ can be quite accurately determined by marker nephelometry.

2. If very accurate measurements of $\overline{\gamma^{*2}}$ are wanted, then spectral data should be obtained into the −1 power-law region predicted by Batchelor [53] so that extrapolation to high wave numbers can be accurately done. It may be noted, though, that attempts in the author's laboratory to verify the −1 power regime experimentally have so far had no success. Further experiments are in progress.

3. The dissipation region of the γ^* spectrum appears to be out of reach experimentally. In the example, $\lambda_\gamma^* = 0.058\ \mu m$, which is around a tenth of the sol particle diameter $D_p = 0.5\ \mu m$. The gas mean free path is $\ell = 0.07\ \mu m$. It may be questioned whether such a small value of λ_γ^* is even meaningful. In any case, pcv's cannot, for several reasons, be made nearly small enough to resolve the far region of the γ^* spectrum when λ_γ^* is so small, e.g., optical limitations on the minimum incident beam diameter, and lack of anything approaching a statistical particle population.

On the basis of the foregoing discussion and the author's experience, the following comments are offered on the measurement of $\overline{\gamma^2}$: if the pcv is small enough to determine $\overline{\gamma^2}$ with reasonable accuracy, then $P = 1$ in (3.4.6). Equations (3.4.6) and (3.4.11) thus give

$$\frac{\overline{\gamma^2}}{\bar{\Gamma}^2} = \frac{\left\{\overline{i_s^{*2}} - \bar{\tau}^2\, \overline{i^2_{msn}} - \overline{i_E^2}\right\}\Big|_{f_1, f_2}}{Q\bar{I}_s^{*2}} \tag{4.2.13}$$

where $\Delta f = f_1 - f_2$ is the effective frequency pass-band for the electrical signal. The low electrical cut-off frequency f_1 should desirably be under $0.1\ \bar{U}/2\pi\Lambda_\gamma$. The high cutoff frequency f_2 should be around $2f_u$ or higher, but desirably no higher than $10f_u$, where $f_u \equiv \kappa_u \bar{U}/2\pi \equiv \bar{U}/2\pi\lambda_u$. The volume-averaging factor Q is virtually unity if $D_V < 0.2\Lambda_\gamma$. When the correction must be made, e.g., if $D_V > 0.2\Lambda_\gamma$, it is recommended that Q be calculated on the assumption that (3.5.7) and (4.2.12) apply at all high wave numbers. Becker *et al.* [20] have shown that then

$$Q \doteq 1 - 0.4\,(D_V/\Lambda_\gamma)^{2/3} \ . \tag{4.2.14}$$

It should be noted, though, that (4.2.12), and hence (4.2.13), are strictly valid only for uniform incident beams; the effect of a laser-type of energy distribution remains to be accurately evaluated. In general, the pcv diameter should be held below $0.5\Lambda_\gamma$ if these procedures are to give the expected high accuracy. It should not be difficult, under reasonably good operating conditions, to determine $\hat{\gamma} \equiv \sqrt{(\overline{\gamma^2})}$ within a probable error of ±5% or better.

It should be remembered that (4.2.14) is based on the assumption of a fairly broad Kolmogoroff equilibrium subrange in which the spectrum follows the (−5/3)-power law. If often happens that while turbulence possesses a spectrum of the form:

$$E^*(\kappa) \propto \overline{\gamma^{*2}}\ \Lambda_\gamma^{1-n}\ \kappa^{-n} \tag{4.2.15}$$

the power of the exponent n is different from 5/3 [20].

Consider the case $n = 2$; then for right-cylindrical pcv's [20]

$$Q = 1 - KD_V/\Lambda_\gamma \tag{4.2.16}$$

where K is an experimental coefficient. The factor Q can also be estimated [20] from the spatial correlation function for concentration fluctuations discussed in Sec.4.3 below. If (4.3.3) applies in the critical region, then for right-cylindrical pcv's with L_i/D_i in the range 0.7÷1.0 [20],

$$Q = 1 - 0.525\, aD_V/\Lambda_\gamma \ , \tag{4.2.17}$$

where $a \equiv a\Lambda_\gamma$. The corresponding result from (4.3.4) is

$$Q = \exp(-0.525\, aD_V/\Lambda_\gamma) \ . \tag{4.2.18}$$

Evidently (4.2.16) and (4.2.17) imply the same forms of the spectrum and correlation functions.

Rosensweig *et al.* [19] considered the form of the correlation function in the limit $\zeta \to 0$,

$$R = 1 - \zeta^2/\lambda_\gamma^{*2} , \tag{4.2.19}$$

where ζ is the separation distance defined in Sec.4.3. This gives

$$Q = 1 - \frac{1}{6}(L_i^2 + \frac{3}{2}D_i^2)/\lambda_\gamma^{*2} . \tag{4.2.20}$$

They also showed that in this case the optimum L_i/D_i ratio is $\frac{1}{2}(3)^{\frac{1}{2}} = 0.866$. The author has indeed normally used L_i/D_i ratios around this value. However, the parabolic behaviour predicted by (4.2.19) as $\zeta \to 0$ is experimentally unobservable, the region being extremely small because of the high Schmidt number of marker particles. Thus the result (4.2.20) is not practically useful, because probes with L_i and D_i around λ_γ^* or smaller are usually impossibly tiny.

4.3 Two-point Spatial Correlation Functions

The preceding discussion of the spectrum $E(\kappa)$ and the mean-square value $\overline{\gamma^2}$ of gas concentration fluctuations should suffice to indicate the general conditions for good measurements and some useful approaches to problems that may be encountered. Space does not permit a detailed examination of every possible type of measurement. One special generalization that is not readily obvious is worth some consideration, however. This concerns the determination of two-point spatial correlation function

$$R(\mathbf{x}_A; \mathbf{x}_B) = \overline{\gamma_A \gamma_B} / \hat{\gamma}_A \hat{\gamma}_B , \tag{4.3.1}$$

where γ_A and γ_B are the gas concentration fluctuations at the two points $\mathbf{x}_A$ and $\mathbf{x}_B$, and $\hat{\gamma} \equiv \sqrt{(\overline{\gamma^2})}$. Becker *et al.* [20] give the example of the symmetrical correlation function $R(x, r, \phi; x, r, \phi + \pi)$ for the statistically axisymmetrical jet considered in Sec.4.3, where $\mathbf{x} \equiv (x, r, \phi)$ in the cylindrical coordinates of the jet. The two pcv's required in the measurement were two segments of the same incident beam, each of length L_i. The distance between the centres of the pcv's is $\zeta \equiv |\mathbf{x}_A - \mathbf{x}_B| = 2r$, and it is expected that $R = R(x, \zeta)$. Theoretically, R should approach $\zeta = 0$ parabolically, and the curvature $\partial^2 R/\partial \zeta^2$ at $\zeta = 0$ and a given x should be related to the rate of dissipation of $\overline{\gamma^2}$,

$$\lambda_\gamma' \equiv \left\{ \frac{2}{\partial^2 R/\partial \zeta^2} \right\}^{\frac{1}{2}} \tag{4.3.2}$$

being the analogue of the Taylor microscale. In the case of a well-developed flow at high turbulence Reynolds number, the parabolic region is however expected to be very small, and virtually insignificant in effect beyond the immediate neighborhood of $\zeta = 0$. The behaviour of R in this parabolic region cannot be determined by marker nephelometry because, as was pointed out in Sec.4.2, the dissipation region of the spectrum of γ is in no wise modelled by the spectrum of γ^*. Generally, the best representation of a curve of $R(x, \zeta)$ *v.* ζ at a given x is to ignore the small region very near $\zeta = 0$ where in fact $\partial R/\partial\zeta \to 0$ and $\partial^2 R/\partial\zeta^2 < 0$, and suppose instead that $\partial R/\partial\zeta < 0$ and $\partial^2 R/\partial\zeta^2 \geqslant 0$ over a large region around $\zeta = 0$. Thus the representations of R as a function of ζ may be typically like

$$R = 1 - a\zeta \tag{4.3.3}$$

or

$$R = \exp(-a\zeta) , \tag{4.3.4}$$

where $a = a(x)$ and $R > 0.2$.

The effect of volume-averaging in the determination of functions such as $R(\mathbf{x},\zeta)$ can be represented by

$$R(\mathbf{x}_A;\mathbf{x}_B) \equiv \frac{\overline{\gamma_A\gamma_B}}{(\overline{\gamma_A{}^2}\,\overline{\gamma_B{}^2})^{1/2}} = \frac{Q_{AB}(\mathbf{x}_A;\mathbf{x}_B)}{(Q_AQ_B)^{1/2}} \frac{\overline{\check{\gamma}_A\check{\gamma}_B}}{(\overline{\check{\gamma}_A{}^2}\,\overline{\check{\gamma}_B{}^2})^{1/2}} . \tag{4.3.5}$$

where $Q_A \equiv Q(\mathbf{x}_A)$, $Q_B \equiv Q(\mathbf{x}_B)$, and $Q_{AB} = Q$ when $\mathbf{x}_A = \mathbf{x}_B$. The volume-averaging factor Q for mean-square concentration fluctuation measurements was discussed in Sec.4.2; we have, for example, (4.2.14) for predicting Q. A rigorous theoretical and experimental study of Q_{AB} has not yet been made; the general procedure would, though, be similar to that for Q as described in [20]. That work in fact suggested an empirical procedure. The results for the case of the symmetrical correlation function $R(x,\zeta)$ in a free jet are shown in Fig.2. It is seen that the data on the volume-averaged function

$$R_V \equiv \overline{\gamma_A{}^2\gamma_B{}^2}/\hat{\gamma}_A\hat{\gamma}_B \tag{4.3.6}$$

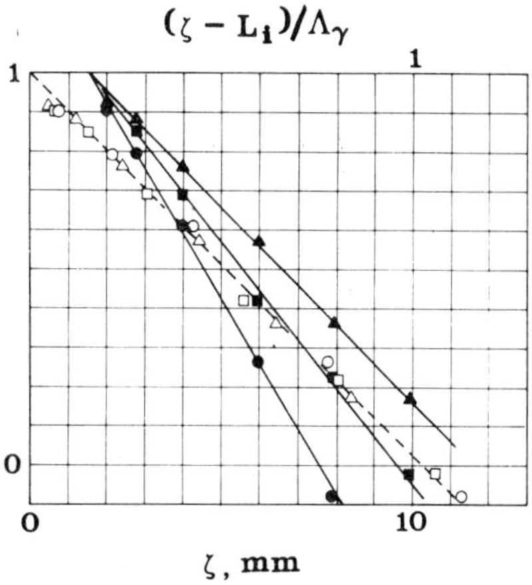

Fig. 2 The symmetrical correlation coefficient as a function of the separation distance ζ in a free jet. Solid curves, the raw data; dashed curve, the data corrected for the displacement error and normalized with respect to the axial integral scale of the concentration fluctuations. The axial downstream distances in nozzle radii are: ○●, $x/r_o = 40$; □■, 56; △▲, 72. (Becker, Hottel and Williams [20], their Fig.7, by permission of *J. Fluid Mech.*).

extrapolate to a common point at $R_V = 1$, and it was noted that at this point $\zeta = L_i$. This suggested that the effect of volume-averaging is essentially a displacement effect, so that

$$R(x, \zeta) = R_V(x, \zeta + \delta) , \qquad (4.3.7)$$

where δ is the displacement distance. In the case considered, where the pcv's were equal segments of the same incident beam, $\delta = L_i$. Figure 2 shows that when the data are plotted as R_V *v.* $(\zeta - L_i)/\Lambda_\gamma$, they are brought together on one curve which behaves in the manner indicated for $R(x, \zeta)$ by (4.3.3) and (4.3.4), where Λ_γ is the longitudinal integral scale given by (4.2.11).

Grandmaison [54] has recently made measurements of $R_V(x, \zeta)$ in an air jet from a 7.6 cm dia. nozzle at sections as far as 3 m downstream of the nozzle; there $L_i \ll \Lambda_\gamma$ and, hence, $R_V = R$. The curves of R *v.* ζ/Λ_γ are in very close agreement with the results of [20] for R_V *v.* $(\zeta - L_i)/\Lambda_\gamma$. The practical validity of (4.3.7) is thereby corroborated.

It is therefore recommended, on the basis of the above results, that two-point correlation functions be corrected for volume averaging by the method indicated by (4.3.7) and Fig.2. The value of the displacement error δ appears to be roughly $\delta = L_i$ when the pcv's are two segments of the same incident beam. For pcv's of other relative geometry, the value of δ will be somewhat different. This procedure should be satisfactory so long as $\Lambda_\gamma \gg \lambda_\gamma$.

A certain implication of the foregoing discussion should be clearly recognized, namely that the determination of turbulence microscales by marker nephelometry is impossible. The temptation to interpret any suggestion of parabolic behaviour around $\zeta = 0$ by means of (4.3.2), and to call the result a microscale, must be firmly rejected.

4.4 *Other Characteristics of Concentration Fluctuations of the Marked Material in Gases*

A variety of other measurements besides those so far considered have been carried out, or are easily possible.

The nephelometer signal can be processed to yield an intermittency factor

$$\bar{\delta} \equiv \text{prob}(\Gamma > 0) ,$$

$$\delta = (1 \text{ if } \Gamma > 0; \;\; 0 \text{ if } \Gamma = 0) . \qquad (4.4.1)$$

Becker, Hottel and Williams [55] describe an electronic system for this operation, and there are several other approaches that can be used such as that described by Professor Libby in the first article in this volume. When $\bar{\delta}$ is known, many of the statistical properties of Γ for those periods when $\Gamma > 0$ can be estimated, as demonstrated in another publication [21]. There are also interesting possibilities in conditional sampling and/or processing; the signal may, for example, be sampled only when $\Gamma > 0$.

The longitudinal two-point correlation function $R(s)$ is generally defined by (4.3.1) with $\mathbf{x}_A$ and $\mathbf{x}_B$ on the same mean-flow streamline, and s the distance between $\mathbf{x}_A$ and $\mathbf{x}_B$ along that streamline. Grandmaison [54] has measured

$R(s)$ along the axis of a round jet; there $\mathbf{x} \equiv (x, r, \phi)$, and for points on the axis

$$\mathbf{x}_A = (x, 0, \text{all } \phi), \qquad \mathbf{x}_B = (x + s,\ 0,\ \text{all } \phi) .$$

The integral longitudinal scale was computed from

$$\Lambda_\gamma \equiv \int_0^\infty R(s)\, ds . \tag{4.4.2}$$

Becker, Hottel and Williams [52] have measured another two-point correlation function in a round jet, namely

$$R(r) \equiv R(\mathbf{x}_A ; \mathbf{x}_B) ,$$

$$\mathbf{x}_A = (x, 0, \text{all } \phi), \qquad \mathbf{x}_B = (x, r, \text{any } \phi) . \tag{4.4.3}$$

The autocorrelation function

$$R(\tau) = \overline{\gamma(t)\gamma(t-\tau)}/\overline{\gamma^2} \tag{4.4.4}$$

can be measured directly, or it can be obtained from the frequency spectrum by means of the Fourier transformation

$$R(\tau) = \frac{1}{\overline{\gamma^2}} \int_0^\infty G(f) \cos(2\pi f t)\, df . \tag{4.4.5}$$

It is supposed in Taylor's hypothesis that $\partial/\partial t = \overline{U}\partial/\partial s$; then $\tau = s/\overline{U}$ and (4.4.5) gives

$$R(s) = \frac{1}{\overline{\gamma^2}} \int_0^\infty E(\kappa) \cos(\kappa s)\, d\kappa , \tag{4.4.6}$$

where $\kappa = 2\pi f / \overline{U}$ and $E(\kappa) = \overline{U} G(f)/2\pi$. Becker *et al.* [21,52] and Grandmaison [54] used this relation to compute $R(s)$ from $E(\kappa)$ for jets, and Grandmaison has compared these results with his direct measurements of $R(s)$ to check Taylor's hypothesis. He found that Λ_γ calculated from (4.4.2) for the direct measurements was 10% lower than the value found through the spectra. Since the turbulence level in jets is high, this level of discrepancy is not surprising.

The inverse of (4.4.6) gives $E(\kappa)$ in terms of $R(s)$, and yields (4.2.11) at $\kappa = 0$,

$$\Lambda_\gamma = \frac{\pi}{2} E(0) / \overline{\gamma^2} . \tag{4.2.11}$$

Becker *et al.* [20,21] have determined Λ_γ on this principle by means of two measurements, namely $\overline{\gamma^2}$ and $\overline{\gamma^2}\big|_{f_1, f_2}$, where f_1 and f_2 are adequately small. The frequency pass-band $\Delta f = f_2 - f_1$ was fixed in respect to f_2 by a low-pass filter with a sharp cutoff characteristic, and in respect to f_1 by the low-frequency limit of the random-signal r.m.s. voltmeter used for the measurements. The effective bandwidth Δf was determined by calibration with white noise. Since the frequency spectrum of γ is expected to be white at very low frequencies (i.e., $G(f)$ is independent of f as $f \to 0$) (4.2.11) indicates that

$$\Lambda_\gamma = \frac{1}{4} U \frac{\overline{\gamma^2}|_{f_1,f_2}}{\overline{\gamma^2}\,\Delta f} \quad . \tag{4.4.7}$$

The high cutoff frequency f_2 should be such that $f_2 < 0.1\, U/2\pi\Lambda_\gamma$, at the very least.

Measurements of the probability density function $f(\gamma)$, or the probability distribution function $F(\gamma)$, are in progress in the author's laboratory. Grandmaison [54] has measured some of the higher moments of the distribution of γ directly by analogue techniques: the skewness $\overline{\gamma^3}/\hat{\gamma}^3$, and the kurtosis $\overline{\gamma^4}/\hat{\gamma}^4$. The radial profile of the kurtosis in a free jet was found to be in remarkably close agreement with the data of Wygnanski and Fiedler [56] on velocity fluctuations, and this is rather to be expected since this quantity is determined primarily by the intermittency characteristics. The skewness, however, differed markedly from that of velocity fluctuations, being generally very much smaller. This indicates that the p.d.f. of concentration fluctuations is much more nearly Gaussian.

Experiments on turbulent diffusion have been carried out by Becker, Rosensweig and Gwozdz [51]. Taylor's theory was applied to determine the Lagrangian integral length scale Λ_L and the intensity of radial velocity fluctuations $\hat{u}_r$, and the results on $\hat{u}_r$ were in good agreement with the data of investigators using hotwire anemometry.

Among higher-order measurements that are quite feasible, but have not yet been attempted, are:

1. Two- and three-point triple covariances, $\overline{\gamma_A{}^2\gamma_B}$ and $\overline{\gamma_A\gamma_B\gamma_C}$.
2. Two-point joint probability density functions $f_{AB}(\gamma_A,\gamma_B)$,

$$f_{AB}(\gamma_A',\gamma_B')d\gamma_A d\gamma_B = \text{prob}(\gamma_A' \leqslant \gamma_A < \gamma_A' + d\gamma_A;\gamma_B' \leqslant \gamma_B < \gamma_B' + d\gamma_B) \quad .$$

There is not much motivation, however, to study such functions until a use is found for them in theory or practice.

5. FEATURES OF FEED CONCENTRATION FIELDS IN GASES

5.1 General

The concentration field of marked material has certain special characteristics that are determined by the number of *physicochemically distinct* feeds entering the system. These characteristics indicate a natural basis for a classification of systems and problems, and an examination of them draws attention to some general properties of processes of mixing and dilation. The qualification "physicochemically distinct" refers essentially to differences in composition and/or temperature. It is supposed in the following discussion that the effects of frictional heating on concentration fields are negligible. When this is not so, frictional heating enters as another source of dilation.

Consider, then, a system entered by a total of n physicochemically distinct (PCD) feeds. The treatment of problems with two or more feeds requires a refinement of the nomenclature used hitherto. In principle, any or all of the feeds to the system can be marked, and it will be taken from here to the end of Sec.6, excepting Sec.5.2, that:

Γ_n = mass concentration of material from the n'th feed,

$\sum_n \Gamma_n = \rho$,

$W_n \equiv \Gamma_n/\rho$ = mass fraction of material from the n'th feed.

In the case of ideal-gas behaviour,

$$\Gamma_n = \rho_{n,o} W_n (M/M_o)(T_o/T) \; . \tag{5.1.1}$$

Similar statements can be made for a given species or complex of atoms from a given feed, e.g.,

$$\Gamma_{C,n} = \Gamma_{C,n,o} W_{C,n}(M/M_o)(T_o/T) \; , \tag{5.1.2}$$

$\Gamma_{C,n,o} \equiv \rho_{n,o} W_{C,n,o}$, but these will usually be taken as implied.

Non-dilational processes pose an interesting subclass of multifeed problems. Non-dilational means without change in volume, thus

$$\nabla \cdot \mathbf{U}^{\ddagger} = 0 \; , \tag{5.1.3}$$

where

$$\mathbf{U}^{\ddagger} \equiv \sum_i \epsilon_i \mathbf{U}_i \equiv \sum_i X_i \tilde{V}_i \mathbf{U}_i$$

is the transport velocity for the volume transport of material, $\epsilon_i = X_i \tilde{V}_i$ is the volume fraction of molecular species i, X_i is the mole fraction of i, $\tilde{V}_i$ is the partial molar volume of i, and $\mathbf{U}_i$ is the transport velocity of i. The velocity $\mathbf{U}^{\ddagger}$ should not be confused with the mass-average velocity – the fluid mechanical stream velocity

$$\mathbf{U} = \sum_i W_i \mathbf{U}_i \; ,$$

where W_i is the mass fraction of i. The statement $\nabla \cdot \mathbf{U} = 0$ implies a constant-density process, whereas a process can be non-dilational without being at constant density, and *vice versa.* Further, $\mathbf{U}^{\ddagger}$ should not be confused with the molar-average velocity

$$\mathbf{U}^{\dagger} = \sum_i X_i \mathbf{U}_i \; ;$$

however, for ideal gases $\tilde{V}_i = \tilde{V} \equiv \sum_i \epsilon_i \tilde{V}_i$ for all i, and then $\epsilon_i = X_i$ and $\mathbf{U}^{\dagger} = \mathbf{U}^{\ddagger}$. The most important example of a non-dilational process is the case of ideal gases mixing at constant temperature and pressure; in this the molar density $c = \rho/M$ is also a constant. A case which may be approximately non-dilational in the process, and is exactly so in the end states, is the mixing of ideal gases fed at different temperatures, but having molar heat capacities that are equal and independent of temperature. In general, in non-dilational gas-phase processes where the ideal gas law applies,

$$\sum_n Y_n = 1 \; , \tag{5.1.4}$$

where $Y_n = \Gamma_n/\rho_{n,o}$, and in the absence of chemical reaction $Y_n = X_n$, where X_n is the mole fraction of material from the nth feed. The mixing of liquids can usually be treated as nondilational.

5.2 The One-feed Problem

Suppose a system is entered by only one PCD feed material. The material may in fact enter by several feed ports (and thus in several streams) but because it is all the same in composition and temperature we can say that the system has only one PCD feed. It will be supposed, for the present argument, that in studying such systems by marker nephelometry all of the feed is uniformly marked. The only process that then produces change in the concentration Γ of marked material is dilation. In the absence of significant pressure changes, this dilation is due to chemical reaction, and the variation in Γ is due to variations in temperature and molecular weight. Here the relation (5.3.1) for ideal-gas behaviour reduces to

$$\Gamma = \Gamma_o(M/M_o)(T_o/T) \tag{5.2.1}$$

where $\Gamma_o = \rho_o$ and $\Gamma = \rho$.

A study of such a problem by marker nephelometry has been carried out by Williams, Hottel and Gurnitz [57]. Their system was a combustion chamber fed with premixed propane and air uniformly marked with magnesium oxide smoke. The turbulent flame was stabilized on a ring flameholder. A statistical characterization of the wrinkled flame front was obtained.

5.3 The Two-feed Problem

Suppose a system is fed with two physicochemically distinct feeds. One of the processes then occurring is invariably mixing. There may be chemical reaction, and the mixing and chemical reaction may be accompanied by dilation. In combustion, one feed may be a fuel gas and the other air. The air may be fed in several streams entering at essentially the same temperature, and which can therefore be taken as a single PCD feed. The fuel may be fed with some premixed air or an inert diluent, and this mixture constitutes one PCD feed. In Sec.5.2, the turbulent premixed flame was given as an example of a one-feed problem. The turbulent diffusion flame, operated without such complications as burnt-gas recirculation, clearly exemplifies two-feed problems in combustion.

One of the PCD feeds may be uniformly marked with sol particles, and the combined process of mixing and dilation can then be studied by marker nephelometry. It is also possible to study the mixing process by means of a physical model as, for example, an air/air mixing model in which appropriate feed air is marked to simulate the process of interest. Nearly all of the investigations hitherto carried out with marker nephelometry, have been of air/air models of two-feed mixing problems.

In two-feed problems

$$\Gamma_1 + \Gamma_2 = \rho \,, \tag{5.3.1}$$

$$W_1 + W_2 = 1 \quad , \tag{5.3.2}$$

and, in nondilational mixing,

$$Y_1 + Y_2 = 1 \quad . \tag{5.3.3}$$

The turbulent fluctuations in these quantities in statistically stationary systems are thus related by

$$\overline{\gamma_1^2} + 2\overline{\gamma_1 \gamma_2} + \overline{\gamma_2^2} = \overline{\rho'^2} \tag{5.3.4}$$

where $\rho' \equiv \rho - \bar{\rho}$ is the density fluctuation;

$$\overline{w_1^2} + 2\overline{w_1 w_2} + \overline{w_2^2} = 0 \quad , \tag{5.3.5}$$

$$\overline{w_1^2} = \overline{w_2^2} , \quad \overline{w_1^3} = -\overline{w_2^3} , \quad \ldots ; \tag{5.3.6}$$

and for nondilational mixing

$$\overline{y_1^2} + 2\overline{y_1 y_2} + \overline{y_2^2} = 0 \ , \tag{5.3.7}$$

$$\overline{y_1^2} = \overline{y_2^2} , \quad \overline{y_1^3} = -\overline{y_2^3} , \quad \ldots \tag{5.3.8}$$

Now $Y_n \equiv \Gamma_n/\Gamma_{n,o} = \Gamma_n/\rho_{n,o}$ is always defined, for dilational as well as non-dilational processes, but (5.3.3), (5.3.7) and (5.3.8) are only valid for non-dilational processes. The quantity Y_n is of prime importance in studies by marker nephelometry; if the nth feed is marked, the transformations from Γ_n^* to Γ_n are generally of the form exemplified by $\Gamma_n/\Gamma_{n,o} = \Gamma_n^*/\Gamma_{n,o}^*$, and thus $Y_n = \Gamma_n^*/\Gamma_{n,o}^*$.

A significant measure of the extent of molecular mixing in the material passing any point in space can be defined as follows. In the absence of mixing by molecular diffusion $(\mathcal{D} = 0)$,

$$\Gamma_1 = \Gamma_{1,o} , \quad \Gamma_2 = 0 \text{ during a fraction of time } \chi;$$

$$\Gamma_1 = 0, \quad \Gamma_2 = \Gamma_{2,o} \text{ during the remaining fraction } 1 - \chi;$$

$$\chi = \bar{\Gamma}_1/\Gamma_{1,o}, \quad 1 - \chi = \bar{\Gamma}_2/\Gamma_{2,o} \ . \tag{5.3.9}$$

These relations give

$$\overline{\gamma_1 \gamma_2} = -\chi(\Gamma_{1,o} + \bar{\Gamma}_1)\bar{\Gamma}_2 - (1-\chi)\bar{\Gamma}_1(\Gamma_{2,o} - \bar{\Gamma}_2) = -\bar{\Gamma}_1\bar{\Gamma}_2 \ . \tag{5.3.10}$$

Then the parameter

$$g_{12} \equiv -\overline{\gamma_1 \gamma_2}/\bar{\Gamma}_1\bar{\Gamma}_2 = -\overline{y_1 y_2}\,/\,\bar{Y}_1\bar{Y}_2 \tag{5.3.11}$$

is an index of molecular segregation, and $1 - g_{12}$ is an index of molecular mixing: g_{12} is unity when the material of the two feeds is completely segregated and zero when it is completely mixed.

It can similarly be shown that $\overline{w_1 w_2} = -\overline{W}_1 \overline{W}_2$ in total segregation; correspondingly we define

$$\omega_{12} \equiv -\overline{w_1 w_2} / \overline{W}_1 \overline{W}_2 \tag{5.3.12}$$

which provides another index of molecular segregation with properties like those of g_{12}.

The value of $\overline{w_1^2}$ in total segregation is

$$\overline{w_1^2} = \chi(1 - \overline{W}_1)^2 + (1 - \chi)\overline{W}_1^2 = \overline{W}_1 (1 - \overline{W}_1) , \tag{5.3.13}$$

and a similar result holds for $\overline{w_2^2}$. Thus a further index of segregation is

$$\omega_n = \overline{w_n^2} / \overline{W}_n (1 - \overline{W}_n) , \tag{5.3.14}$$

$n = 1$ or 2, Here, of course, $\overline{w_1^2} = \overline{w_2^2} = -\overline{w_1 w_2}$, $\overline{W}_n (1 - \overline{W}_n) = \overline{W}_1 \overline{W}_2$, and so $\omega_n = \omega_{12}$. It will be seen in Sec.5.4, however, that (5.3.12) and (5.3.13) are each applicable to problems with more than two feeds, but there $\omega_1 \neq \omega_2 \neq \omega_{12}$.

In nondilational processes $\overline{y_n^2} = \overline{Y}_n(1 - \overline{Y}_n)$, and the analogue of (5.3.14) is

$$g_n' = \overline{y_n^2} / \overline{Y}_n (1 - \overline{Y}_n) , \tag{5.3.15}$$

where the prime superscript is a reminder of the restriction on the conditions under which $g_n' = g_{12}$.

The idea that quantities of the form of (5.3.14) or (5.3.15) are measures of segregation in two-feed problems was introduced into mixing theory by Danckwerts [58] who gave them the name *intensity of segregation.* The idea that concentration fluctuations are associated with incomplete mixing was suggested earlier by Hawthorne, Weddell and Hottel [59] who spoke of quantities such as $\hat{\gamma}_n$ as *unmixedness factors.*

Correlation functions such as

$$R_{12} \equiv \overline{\gamma_1 \gamma_2} / \hat{\gamma}_1 \hat{\gamma}_2 = \overline{y_1 y_2} / \hat{y}_1 \hat{y}_2 \tag{5.3.16}$$

$$\Omega_{12} \equiv \overline{w_1 w_2} / \hat{w}_1 \hat{w}_2 \tag{5.3.17}$$

are not indicators of segregation in two-feed problems. In nondilational processes, the quantity Ω_{12} always equals -1 while R_{12} does so too in nondilational processes. In general, from (5.3.4),

$$\overline{\gamma_1 \gamma_2} = \frac{1}{2}(\overline{\rho'^2} - \overline{\gamma_1^2} - \overline{\gamma_2^2}) . \tag{5.3.18}$$

Hence $1 + R_{12}$ indicates effects of dilation.

There have been no experimental investigations of the quantities in (5.3.4) for dilational processes. All that is necessary to do this by marker nephelometry is to mark first one stream, then the other, and finally both, and thus obtain $\overline{\gamma_1^2}$, $\overline{\gamma_2^2}$ and $\overline{\rho'^2}$. The value of $\overline{\gamma_1\gamma_2}$ is then found from (5.3.4).

5.4 *The Three-feed Problem*

In the three-feed problem

$$\Gamma_1 + \Gamma_2 + \Gamma_3 = \rho \,, \tag{5.4.1}$$

$$W_1 + W_2 + W_3 = 1 \,. \tag{5.4.2}$$

In statistically stationary systems

$$\overline{\gamma_1^2} + \overline{\gamma_2^2} + \overline{\gamma_3^2} + 2\overline{\gamma_1\gamma_2} + 2\overline{\gamma_2\gamma_3} + 2\overline{\gamma_1\gamma_3} = \overline{\rho'^2} \,, \tag{5.4.3}$$

$$\overline{w_1^2} + \overline{w_2^2} + \overline{w_3^2} + 2\overline{w_1w_2} + 2\overline{w_2w_3} + 2\overline{w_1w_3} = 0 \,, \tag{5.4.4}$$

$$\overline{w_n^2} = \overline{w_l^2} + 2\overline{w_lw_m} + \overline{w_m^2} \,, \tag{5.4.5}$$

where l, m and n take values of 1, 2, 3, and $l \neq m \neq n$. The relations for Y_n and y_n, $n = 1, 2, 3$ for nondilational processes are analogous to (5.4.2), (5.4.4) and (5.4.5).

Indices of segregation for pairs of feeds are defined as in the two-feed problem:

$$g_{lm} = \overline{\gamma_l\gamma_m}/\overline{\Gamma}_l\overline{\Gamma}_m = -\overline{y_ly_m}/\overline{Y}_l\overline{Y}_m \,, \tag{5.4.6}$$

$$\omega_{lm} = -\overline{w_lw_m}/\overline{W}_l\overline{W}_m \,. \tag{5.4.7}$$

Consider the situation when the feeds of l and m are totally segregated, but each may be mixed to some degree with the third feed, n:

$$\Gamma_l = \overline{\Gamma}_l' + \gamma_l' \,, \quad \Gamma_m = \overline{\Gamma}_m \quad \text{during a fraction of time } \chi,$$

$$\Gamma_l = \overline{\Gamma}_l \,, \quad \Gamma_m = \overline{\Gamma}_m' + \gamma_m' \quad \text{during the remaining fraction } 1-\chi,$$

$$\Gamma_l = \overline{\Gamma}_l'/\chi \,, \quad \overline{\Gamma}_m = \overline{\Gamma}_m'/(1-\chi) \,, \tag{5.4.8}$$

where

$\overline{\Gamma}_l'$ = mean value of Γ_l for that fraction of time during which

$\Gamma_l > 0$,

$\overline{\Gamma}_m'$ = mean value of Γ_m for that fraction of time during which

$$\Gamma_m > 0 \ ,$$

and $\gamma_l' \equiv \Gamma_l - \bar{\Gamma}_l$, $\gamma_m' \equiv \Gamma_m - \bar{\Gamma}_m'$ are turbulent fluctuations due to mixing with the third feed. Then

$$\overline{\gamma_l\gamma_m} = -\chi\langle\bar{\Gamma}_m(\bar{\Gamma}_l'+\gamma_l'-\bar{\Gamma}_l)\rangle_\chi - (1-\chi)\langle\bar{\Gamma}_l(\bar{\Gamma}_m'-\gamma_m'-\bar{\Gamma}_m)\rangle_{1-\chi} \tag{5.4.9}$$

where the first average is over the fraction of time when $\Gamma_l > 0$, $\Gamma_m = 0$, and the second over that fraction when $\Gamma_l = 0$, $\Gamma_m > 0$. The final result is $\overline{\gamma_l\gamma_m} = -\bar{\Gamma}_l\bar{\Gamma}_m$. By the same argument, $\overline{w_l w_m} = -\bar{W}_l\bar{W}_m$. Thus $g_{lm} = \omega_{lm} = 1$ when the feeds l and m are totally segregated, just as in the two-feed problem.

When the material of all three feeds is segregated,

$$W_n = 1 \text{ during a fraction of time } \chi_n,$$

$$W_n = 0 \text{ during the remaining fraction } 1-\chi_n,$$

and $\bar{W}_n = \chi_n$. Thus

$$\overline{w_n^2} = \chi_n(1-\bar{W}_n)^2 + (1-\chi_n)\bar{W}_n^2 = \bar{W}_n(1-\bar{W}_n) \ . \tag{5.4.10}$$

This result, already encountered in (5.3.13), applies to any multifeed problem wherever all the feed materials are totally segregated. The index of segregation ω_n defined by (5.3.14) is therefore generally relevant. The analogous result for non-dilational processes in terms of γ_n, namely (5.3.15), also applies.

A further result for the state when all three feed materials are segregated is

$$\overline{\gamma_1\gamma_2\gamma_3} = \chi_1(\Gamma_{1,o}-\bar{\Gamma}_1)\bar{\Gamma}_2\bar{\Gamma}_3 + \chi_2\bar{\Gamma}_1(\Gamma_{2,o}-\bar{\Gamma}_2)\bar{\Gamma}_3 + \chi_3\bar{\Gamma}_1\bar{\Gamma}_2(\Gamma_{3,o}-\bar{\Gamma}_3) \ , \tag{5.4.11}$$

$$\bar{\Gamma}_n = \chi_n\Gamma_{n,o} \ , \quad \sum_n \chi_n = 1 \ ,$$

and so

$$\overline{\gamma_1\gamma_2\gamma_3} = \bar{\Gamma}_1\bar{\Gamma}_2\bar{\Gamma}_3 \ . \tag{5.4.12}$$

It might thus be thought that

$$g_{123} = \overline{\gamma_1\gamma_2\gamma_3}/\bar{\Gamma}_1\bar{\Gamma}_2\bar{\Gamma}_3 = \overline{y_1y_2y_3}/\bar{Y}_1\bar{Y}_2\bar{Y}_3 \tag{5.4.13}$$

is an index of segregation for the three feed materials together.

Now, at all stages of mixing, $w_1 + w_2 + w_3 = 0$. Thus

$$\overline{w_1w_2w_3} = -\overline{w_l w_m^2} - \overline{w_l^2 w_m} \ , \tag{5.4.14}$$

where l and m are 1, 2 or 3 and $l \neq m$. Evaluation of

$$\overline{(w_1 + w_2 + w_3)^3} = 0$$

then shows that the skewnesses of w_1, w_2 and w_3 are related by

$$\overline{w_1^3} + \overline{w_2^3} + \overline{w_3^3} = 0 \,. \tag{5.4.15}$$

Again, because $w_1 + w_2 + w_3 = 0$,

$$\overline{(w_1 + w_2)^3} = \overline{w_1^3} + 3\overline{w_1^2 w_2} + 3\overline{w_1 w_2^2} + \overline{w_2^3} = -\overline{w_3^3} \,. \tag{5.4.16}$$

Then (5.4.14) and (5.4.16) give

$$\overline{w_1 w_2 w_3} = 0 \,. \tag{5.4.17}$$

Expressions analogous to (5.4.14–17) apply to y_1, y_2 and y_3 in nondilational mixing. It follows that the values of g_{123} is associated with dilation, and $g_{123} = 0$ in nondilational mixing.

It is possible that γ_l and γ_m, or w_l and w_m, may be positively correlated in the more advanced stages of mixing. Hence $g_{lm} = 0$ or $\omega_{lm} = 0$ does not generally mean that the materials l and m are perfectly mixed. A better indication of the completion of mixing may here be given by the double correlation functions

$$R_{lm} \equiv \overline{\gamma_l \gamma_m} / \hat{\gamma}_l \hat{\gamma} \equiv \overline{y_l y_m} / \hat{y}_l \hat{y}_m \,, \tag{5.4.18}$$

$$\Omega_{lm} = \overline{w_l w_m} / \hat{w}_l \hat{w}_m \,. \tag{5.4.19}$$

The possible range of these functions is from -1 to $+1$, but their behaviour is considerably dependent on the characteristics of each system.

In studying a three-feed problem by marker nephelometry, the information of greatest interest can often be obtained by marking only two of the feeds. Suppose feed 1 is first marked, then feed 2, and finally both 1 and 2. The sum of the fields of 1 and 2 is

$$\Gamma_1 + \Gamma_2 = \Gamma_3' \,, \tag{5.4.20}$$

$\Gamma_3' = \rho - \Gamma_3$. Then

$$\gamma_1 + \gamma_2 = \gamma_3' = -\gamma_3 \tag{5.4.21}$$

and

$$\overline{\gamma_1^2} + 2\overline{\gamma_1 \gamma_2} + \overline{\gamma_2^2} = \overline{\gamma_3'^2} = \overline{\gamma_3^2} \,. \tag{5.4.22}$$

Three-feed problems constitute an important class. In combustion, one of the feeds may be fuel, the second air, and the third may be oxygen, or an inert, or recirculated combustion gases. An experimental study of a significant cold mixing model has been made by Becker and Booth [60]. Their system was of two interacting free air jets mixing with each other and a virtually infinite atmosphere of room air. The angle between the intersecting axes of the round nozzles was varied between 30° and 90°. The one nozzle air stream was first marked, then the other, and finally both. The fields of $\overline{Y}_1$, $\overline{Y}_2$, $\overline{Y'_3}$, $\overline{y_1^2}$, $\overline{y_2^2}$, and $\overline{y_3'^2}$ were mapped, and from these results R_{12} as given by (5.4.18) was computed. The field of R_{12} gave a very revealing picture of the mixing between the jet feeds in the jet interaction zone. An example is given in Fig.3, from which it is seen that R_{12} went

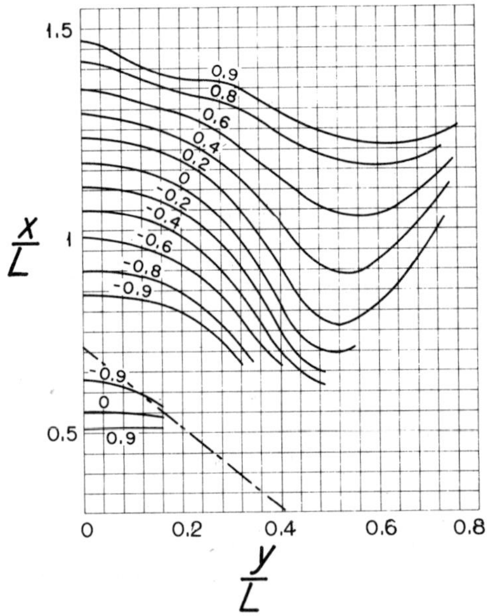

Fig.3 Field of the correlation function $R_{12} \equiv \overline{\gamma_1 \gamma_2}/\hat{\gamma}_1 \hat{\gamma}_2$ for two interacting round jets whose nozzle axes intersect at a half-angle of 45°. The dash-dotted line indicates the nozzle axes (Becker and Booth [60], their Fig.12, by permission of *A. I. Ch. E. Jour.* Note: Figs 3–9 and 12 in the reference are in error in that the transverse position y/L should be multiplied by 2 and the indicated position of the nozzle axes should be shifted accordingly. Readers can obtain corrected figures with the supplementary material mentioned at the end of the article. The figure here reproduced shows corrected results.)

from near -1 at the first contact of the jets to approach $+1$ asymptotically at the end. The significance of $R_{12} = +1$ in this case is that downstream of the interaction zone the jet feeds are perfectly mixed, but the combined jet continues to entrain ambient air (the third feed) and this is a continuing source of concentration fluctuations. The fluctuations γ_1 and γ_2 are however perfectly correlated, and thus $R_{12} = 1$ in this region.

To see what these results might mean in a combustion situation, suppose that the one jet feed is air, the second is fuel gas, and the ambient gas is combustion products. Suppose further that the combustion process is diffusion-controlled. The progress of R_{12} towards +1 then follows the progress of combustion, and the close approach of R_{12} to +1 indicates the tip of the flame.

It is evident that the fields of correlation functions such as R_{12} can be of basic significance for three-feed problems, particularly with regard to chemical reaction. There are many other possible studies, besides the one just described, that would be useful in relation to combustion theory and practice. An investigation is now in progress in the author's laboratory of a system of two opposed jets injected transversely into a pipe flow.

5.5 Problems with More Than Three Feeds

From the preceding discussion, the segregation indices for pairs of feed materials, namely g_{lm} and ω_{lm}, are valid for all multistream problems; so too are the segregation intensities ω_n and g_n' and the correlation functions R_{lm} and Ω_{lm}. Useful studies can be carried out by marker nephelometry by focussing attention on pairs of feeds, marking first one, then the other, and finally both.

The behaviour of higher covariances such as $\overline{\gamma_1\gamma_2\gamma_3}$, $\overline{\gamma_1\gamma_2^2}$, $\overline{\gamma_1\gamma_2\gamma_3\gamma_4}$, etc. is, however, quite complex; measurements would be difficult or impossible, and most of these quantities have no obvious physical significance.

6. THEORY OF MATHEMATICAL AND PHYSICAL MODELLING FOR GASES

6.1 Governing Equations for Material Transport

The equation of transport of a molecular species i is generally

$$\rho DW_i/Dt + \nabla\cdot \mathbf{j}_i + r_i = 0 \; , \tag{6.1.1}$$

where $\mathbf{j}_i \equiv \rho W_i(\mathbf{U}_i - \mathbf{U})$ is the diffusion mass flux vector, and r_i is the mass rate of homogeneous chemical reaction per unit volume. If only ordinary diffusion is important*, if the binary diffusivities for all pairs of molecules – then denoted by $\mathcal{D}$ – are approximately equal, and if the local variation of the mixture molecular weight is small, the Stefan-Maxwell equations give [49]

$$\mathbf{j}_i = -\mathcal{D}\rho\nabla X_i = -\mathcal{D}\rho\nabla W_i \; . \tag{6.1.2}$$

Then

$$\rho DW_i/Dt - \nabla\cdot\rho\mathcal{D}\nabla W_i + r_i = 0 \; . \tag{6.1.3}$$

The conventional approach to equations such as (6.1.1) for statistically stationary turbulent systems has been to apply Reynolds averaging – i.e., to take the simple time-average of each quantity, so that for ρQ

---

* The other types of diffusion are thermal diffusion (due to temperature gradients), pressure diffusion (due to pressure gradients), and forced diffusion (due to gravitational, electrostatic and magnetic fields).

$$<\rho Q>_{\text{Re}} \equiv \overline{\rho Q} = \lim_{T \to \infty} \frac{1}{T} \int_t^{t+T} \rho Q \, dt \ .$$

Favre [61] has suggested that for variable-density flows it is advantageous to use a different averaging process such that

$$\overline{\rho Q} = \bar{\rho} <Q>_{\text{Fa}} \ .$$

The expansion of a Reynolds average

$$\overline{\rho Q} = \bar{\rho}\,\bar{Q} + \overline{\rho' q} \ ,$$

where $Q = \bar{Q} + q$, involves covariances $\overline{\rho' q}$ with density fluctuations, whereas Favre averages do not. Bilger [62] has applied Favre averaging to combustion problems and, like Professor Libby in his companion article in this volume, strongly advocates its adoption. It is meaningful here to write $r_i \equiv \rho Z_i$; the Favre average of (6.1.2) is then

$$\nabla \cdot \bar{\rho} <W_i U>_{\text{Fa}} - \nabla \cdot \bar{\rho} <\mathcal{D} \nabla W_i>_{\text{Fa}} + \bar{\rho} <Z_i>_{\text{Fa}} = 0 \ . \qquad (6.1.4)$$

Further discussion of variable-density turbulent flows is contained in Professor Libby's article.

In the case of constant-density flows, the Favre and Reynolds averages are identical. The continuity equation

$$\partial \rho / \partial t + \nabla \cdot \rho \mathbf{U} = 0 \qquad (6.1.5)$$

then reduces to $\nabla \cdot \mathbf{U} = 0$. If, further, $\mathcal{D}$ is a constant, then (6.1.4) can be written in the familiar Reynolds-averaged form

$$\bar{\mathbf{U}} \cdot \nabla \bar{W}_i + \nabla \cdot \overline{w_i \mathbf{u}} - \mathcal{D} \nabla^2 \bar{W}_i + \bar{Z}_i = 0 \ . \qquad (6.1.6)$$

This equation, and all the equations given hereinafter for constant-density flows, can obviously be written in terms of mass concentration instead of mass fraction by simply substituting the one for the other; thus

$$\bar{\mathbf{U}} \cdot \nabla \bar{\Gamma}_i + \nabla \cdot \overline{\gamma_i \mathbf{u}} - \mathcal{D} \nabla^2 \bar{\Gamma}_i + \rho \bar{Z}_i = 0 \ . \qquad (6.1.7)$$

The mass fraction of an element k in a molecule i is β_{ik}, and the mass fraction W_k of k in a molecular mixture is

$$W_k = \sum_i \beta_{ik} W_i \ . \qquad (6.1.8)$$

In the case where the binary diffusivities of all significant molecular pairs are essentially equal and (6.1.2) applies, (6.1.1) gives

$$\rho DW_k/Dt - \nabla\cdot\rho\mathcal{D}\nabla W_k = 0 \ , \qquad (6.1.9)$$

the chemical reaction effect being nil because atoms are conserved. A similar equation also applies to any distinct complex of elements, for example all the C and H atoms present, or all the atoms originating in the nth feed material, or all the C and H atoms introduced in the nth feed material. Thus, for the nth feed,

$$\rho DW_n/Dt - \nabla\cdot\rho\mathcal{D}\nabla W_n = 0. \qquad (6.1.10)$$

For statistically stationary constant-density turbulent flows with $\mathcal{D}$ constant, (6.1.8) gives

$$\overline{\mathbf{U}}\cdot\nabla\overline{W}_k + \nabla\cdot\overline{w_k\mathbf{u}} - \mathcal{D}\nabla^2\overline{W}_k = 0 \ , \qquad (6.1.11)$$

while (6.1.10) gives

$$\overline{\mathbf{U}}\cdot\nabla\overline{W}_n + \nabla\cdot\overline{w_n\mathbf{u}} - \mathcal{D}\nabla^2\overline{W}_n = 0 \ . \qquad (6.1.12)$$

Equation (6.1.1) can also be written with Γ_k and γ_k substituted for W_k and w_k, and (6.1.12) with Γ_n and γ_n, or Y_n and y_n, substituted for W_n and w_n.

In the special case of a two-feed nondilational process in an ideal gas system without chemical reactions, it is easily shown that

$$\frac{DY_1}{Dt} - \nabla\cdot\mathcal{D}\left[1-(1-M_1/M_2)Y_1\right]^{-1}\nabla Y_1 = 0 \qquad (6.1.13)$$

where subscripts 1 and 2 denote the feed materials. Then in steady turbulent flow

$$\overline{\mathbf{U}}\cdot\nabla\overline{Y}_1 + \nabla\cdot\overline{y_1\mathbf{u}} - \nabla\cdot\mathcal{D}\left[1-(1-M_1/M_2)\overline{Y}_1\right]^{-1}\nabla\overline{Y}_1 = 0 \qquad (6.1.14)$$

approximately.

Bilger [63,64] has derived an interesting result for the chemical reaction rate r_i. Suppose the system is such that we may suppose that $W_i = W_i(W_n)$; then

$$\frac{dW_i}{dt} = \frac{dW_i}{dW_n}\frac{dW_n}{dt} \ . \qquad (6.1.15)$$

From (6.1.3) and (6.1.10),

$$r_i = -\rho\frac{DW_i}{Dt} + \nabla\cdot\rho\mathcal{D}\nabla W_i$$

$$= -\frac{dW_i}{dW_n}\frac{DW_n}{Dt} + \nabla\cdot\rho\mathcal{D}\frac{dW_i}{dW_n}\nabla W_n$$

$$= -\frac{dW_i}{dW_n} \nabla\cdot\rho\mathcal{D}\nabla W_n + \nabla\cdot\rho\mathcal{D}\frac{dW_i}{dW_n}\nabla W_n$$

$$= \rho\mathcal{D}\nabla W_n\cdot\nabla\left\{\frac{dW_i}{dW_n}\right\}$$

$$= \rho\mathcal{D}\frac{d^2W_i}{dW_n^{\,2}}\nabla W_n\cdot\nabla W_n \qquad (6.1.16)$$

or

$$Z_i = \mathcal{D}\frac{d^2W_i}{dW_n^{\,2}}\nabla W_n\cdot\nabla W_n \ . \qquad (6.1.17)$$

The implications of the condition $W_i = W_i(W_n)$ are not difficult to see for a two-feed system; either (i) the chemical reactions are all fast relative to the turbulence time scale, and one-way, or (ii) the reactions are all fast and the system is at every point in thermodynamic equilibrium. A typical example of the first case is the assumption, sometimes made for a diffusion flame, that the fuel (a hydrocarbon, say) is fully reacted to CO_2 and H_2O at every point to the extent that O_2 is available in the mixture. An example of the second case is discussed in Sec.6.4. In a two-feed system $n = 1$ or 2, and W_1 and W_2 are uniquely related by $W_1 + W_2 = 1$. Thus the elemental composition is completely determined at every point, for given feed compositions, by the value of either W_1 or W_2.

Transport equations can also be written for the covariances of composition or concentration fluctuations, and some of these are required in the more advanced computational schemes for turbulent flows. For molecular species i and j, the equation of the double covariance for a constant-density flow under the conditions of (6.1.2) with $\mathcal{D}$ constant is [60]

$$\bar{\mathbf{U}}\cdot\nabla\overline{w_iw_j} + \overline{w_j\mathbf{u}}\cdot\nabla\bar{W}_i + \overline{w_i\mathbf{u}}\cdot\nabla\bar{W}_j + \nabla\cdot\overline{w_iw_j\mathbf{u}} - \mathcal{D}\nabla^2\overline{w_iw_j}$$
$$+ 2\mathcal{D}\overline{\nabla w_i\cdot\nabla w_j} + \overline{w_j\zeta_i} + \overline{w_i\zeta_j} = 0 \ , \qquad (6.1.18)$$

where $\zeta_i \equiv Z_i - \bar{Z}_i$, $\zeta_j \equiv Z_j - \bar{Z}_j$. The first term is identified with convection of $\overline{w_iw_j}$, the second and third terms with production, the fourth with transport by turbulent diffusion, the fifth with transport by molecular diffusion, the sixth with dissipation by molecular diffusion, and the last two terms with dissipation by chemical reaction. When $i = j$, we obtain the equation of the mean-square mass fraction fluctuation $\overline{w_i^2}$,

$$2\overline{w_i\mathbf{u}}\cdot\nabla\bar{W}_i + \bar{\mathbf{U}}\cdot\nabla\overline{w_i^2} + \nabla\cdot\overline{w_i^2\mathbf{u}} - \mathcal{D}\nabla^2\overline{w_i^2} + 2\mathcal{D}\overline{\nabla w_i\cdot\nabla w_i} + 2\overline{w_i\zeta_i} = 0 \ . \qquad (6.1.19)$$

Equations similar to (6.1.18) and (6.1.19), and subject to the same conditions, can be written for any element k, complexes of elements, and the material of any given feed. For any two feeds l and m, for example,

$$\overline{\mathbf{U}}\cdot\nabla\overline{w_l w_m} + \overline{w_l\mathbf{u}}\cdot\nabla\overline{W}_m + \overline{w_m\mathbf{u}}\cdot\nabla\overline{W}_l + \nabla\cdot\overline{w_l w_m \mathbf{u}} - \mathcal{D}\nabla^2\,\overline{w_l w_m} + 2\mathcal{D}\overline{\nabla w_l\cdot\nabla w_m} = 0\,, \tag{6.1.20}$$

and this can also be written with γ_l and γ_m, or y_l and y_m, in place of W_l and w_m.

The double covariance $\overline{\gamma_i\gamma_j}$ of fluctuations in molecular mass concentration arises in the formulation of bimolecular chemical reactions, and (6.1.19) is pertinent in this context. The double covariance of the mass concentrations of feed materials $\overline{\gamma_l\gamma_m}$ is significant, as shown earlier, in assessing the degree of mixing in problems involving three or more feeds, through $R_{lm} \equiv \overline{\gamma_l\gamma_m}/\hat{\gamma}_l\hat{\gamma}_m$. The value of $\overline{\gamma_l\gamma_m}$ may be relevant to chemical reaction when, for instance, molecular species i is introduced with feed l, and species j with feed m; in the case of a bimolecular reaction $\overline{y_l y_m}$ may indicate the maximum possible value of $\overline{y_i y_j}$ – that which would pertain in the absence of chemical reaction. The values of $\overline{\gamma_l\gamma_m}$ and R_{lm} are also of interest for three-feed problems when the rate of chemical reaction is essentially diffusion-controlled.

Equations can also be developed for triple covariances such as $\overline{\gamma_l\gamma_m\gamma_n}$, and for even higher covariances, but the usefulness of these remains to be established. It has been noted that in three-feed problems $\overline{w_1 w_2 w_3} = 0$ generally, and $\overline{\gamma_1\gamma_2\gamma_3} = 0$ in nondilational mixing. When these quantities are non-zero, e.g. $\overline{\gamma_1\gamma_2\gamma_3}$ in a four-feed problem, it does not appear that sufficiently accurate measurements can be made to determine them.

For instance, the determination of $\overline{y_1 y_2 y_3}$ by marker nephelometry in a four-feed problem of constant-density mixing requires measurements with feeds 1, 2 and 3 individually marked, marked in pairs, and marked altogether. The last marking gives the value of

$$\overline{(y_1 + y_2 + y_3)^3} = \overline{y_4'^3} = \overline{y_4^3}\,, \tag{6.1.21}$$

where $y_4' = \gamma_4'/\rho = y_1 + y_2 + y_3$ is the fluctuation when the three given feeds are all marked. The l.h.s. of (6.1.22) involves the covariances $\overline{y_l y_m^2}$ in addition to $\overline{y_1 y_2 y_3}$, and these must be determined from $\overline{(y_l + y_m)^3}$. The net result is that $\overline{y_1^3}$, $\overline{y_2^3}$, $\overline{y_3^3}$ and $\overline{y_3'^3}$ must all be accurately measured, and then $\overline{y_1 y_2 y_3}$ can be computed.

6.2 *Governing Equations for Momentum and Energy Transport*

The fluid-mechanical governing equations – the equations for $\overline{\mathbf{U}}$, $e_K \equiv \frac{1}{2}\overline{\mathbf{u}\cdot\mathbf{u}}$, ϵ_T, etc., are required in any complete solution scheme, but a discussion of them is outside the scope of the present article. In any case, these equations have been amply presented by other authors [e.g., 41–43, 63, 65–69].

Also required is the equation of conservation of energy, usually best expressed as an enthalpy transport equation. The Schwab-Zeldovich formulation, discussed by Williams [70], is appropriate under the present simplifying assumptions for gaseous systems – those given for (6.1.2) – with the additional assumption that the Lewis number is unity, and thus $a \equiv \kappa/\rho C_p = \mathcal{D}$. The total enthalpy form is usually most convenient [49]; it is

$$\rho Dh/Dt - \nabla\cdot\mathcal{D}\rho\nabla h = 0 \tag{6.2.1}$$

where

$$h = \sum_i W_i(h_i^0 + \int_{T^\circ}^{T} C_{pi}\,dT) \ . \tag{6.2.2}$$

Becker [49], for instance, has considered the two-feed problem of a free-jet turbulent diffusion flame of a fuel gas burning in air; for this case, under the given assumptions,

$$W_1/W_{1,0} = (h - h_{2,0})/(h_{1,0} - h_{2,0}) \ , \tag{6.2.3}$$

where 1 is the fuel feed and 2 is the air feed.

6.3 *Turbulence Models*

Some or all of the Reynolds- or Favre-averaged transport ("governing") equations for a turbulent flow contain various covariances between turbulently fluctuating variables. Although a transport equation might be introduced for each of these covariances, this is not done in practice beyond a certain level of complexity, in order to effect closure of the system of equations. The covariances not described by individual transport equations are then represented by special relations which are equivalent in function to the constitutive equations or laws used in molecular transport theory, but which here are usually called turbulence models or hypotheses. The problems of turbulence modelling have been examined by many authors [e.g., 42, 63–72]. A general review is not attempted here, and the following discussion will be limited to a brief consideration of interesting terms in the material transport equations for constant-density flows.

In the transport equation for $\overline{W}_i$, (6.1.6), the turbulent flux $\overline{w_i\mathbf{u}}$ is usually represented by the gradient diffusion model

$$\overline{w_i\mathbf{u}} = -C_1\nu_T\nabla\overline{W}_i \tag{6.3.1}$$

where ν_T is the kinematic turbulence viscosity and C_1 is the reciprocal of the turbulence Schmidt number. The value of C_1 is usually assumed to be the same in the presence of chemical reaction as without. An analogous expression can be written for the nth feed material,

$$\overline{w_n\mathbf{u}} = -C_1\nu_T\nabla\overline{W}_n \ . \tag{6.3.2}$$

The transport equation for $\overline{w_iw_j}$, (6.1.19), involves the turbulent flux density $\overline{w_iw_j\mathbf{u}}$. This also may be represented by a gradient diffusion model

$$\overline{w_iw_j\mathbf{u}} = -C_2\nu_T\nabla\overline{w_iw_j} \ , \tag{6.3.3}$$

both for $i = j$ and $i \neq j$. An analogous result applies to $\overline{w_l w_n \mathbf{u}}$ in (6.1.21).

The equation of $\overline{w_i w_j}$ also contains the dissipation term $\mathcal{D}\nabla w_i \cdot \nabla w_j$. It may here be hypothesized that

$$\overline{\nabla w_i \cdot \nabla w_j} = C_3 \epsilon_T \, \overline{w_i w_j} / \nu e_K \ , \tag{6.3.4}$$

both for $i = j$ and $i \neq j$. An analogous expression applies to $\overline{\nabla w_l \cdot \nabla w_n}$ in (6.1.21).

Finally, consider Bilger's chemical rate expression for the case $W_i = W_i(W_n)$, (6.1.18). In a field with strong turbulent concentration fluctuations this approximates to

$$\overline{Z}_i = \mathcal{D} \frac{d^2 \overline{W}_i}{d\overline{W}_n^2} \overline{\nabla w_n \cdot \nabla w_n} \ , \tag{6.3.5}$$

and thus

$$\overline{Z}_i = C_4 \frac{\mathcal{D} \epsilon_T \overline{w_i w_j}}{\nu e_K} \frac{d^2 \overline{W}_i}{d\overline{W}_n^2} \ . \tag{6.3.6}$$

In the above results, γ_i can be substituted for w_i, and γ_n or y_n for w_n, the forms for constant density flows all being alike. It should be understood that the turbulent viscosity ν_T may itself be represented by, e.g., $\nu_T = \Lambda e_K{}^{1/2}$, where Λ is a suitably defined integral length scale.

The study of systems by marker nephelometry can provide data from which the constants in models such as (6.3.1) – (6.3.4) may be estimated. The data can also be used to check the performance of full simulations – i.e., the numerical solutions of the governing equations under given boundary conditions and turbulence models. Spalding [71] was perhaps the first to have predicted the field of concentration fluctuations for a turbulent free jet; his calculations demonstrated close agreement with the results of Becker *et al.* [21]. Dr. Elghobashi, in the article following this one, provides further examples of computing confined jet behaviour including flows with recirculation.

6.4 *Physical Modelling of Turbulent Diffusion Flames*

Physical modelling – e.g., air/air mixing studies of two-feed systems – can provide much useful information about a system's performance, and can partly or wholly substitute for a full mathematical simulation. In some cases the interpretation of a physical model is very straightforward. In others there is necessity for a considerable play of intuition and hypothesis. The latter is certainly the case in the cold modelling of combustion systems, and two examples will now be examined to illustrate approaches to such difficult situations.

Becker, Hottel and Williams [72] mapped the fields of mean velocity and mean concentration for an axisymmetrical confined jet system for various values of the confined-jet similarity parameter Ct – the Craya-Curtet number. It is possible to interpret this study as a cold simulation of a tunnel furnace fired from one end with a central jet of hydrocarbon fuel gas and a surrounding uniform flow of air, and to estimate the radiant heat flux to the tunnel walls on the assumption that luminous radiation is unimportant relative to gas radiation. Assume, for example, that:

1. The field of the mass flux density $\bar{\rho}\,\bar{U}_x$ is the same as in the cold system, $\bar{\rho}\bar{U}_x = \bar{\rho}'\bar{U}_x'$, where the primes indicate the model values and U_x is the axial component of velocity.

2. The profiles of $\bar{W}_1$ in the hot system are the same as the profiles of $\bar{Y}_1$ in the cold system, thus $\bar{W}_1 = \bar{Y}_1'$, where subscript 1 denotes the fuel material.

3. Fuel and oxygen are everywhere fully reacted to CO_2 and H_2O to the extent that oxygen is available.

Sarofim and Hottel [73] have carried out computations from a model like the above, using the cited data of Becker *et al.* The results provide a revealing picture of the effect of including real concentration and velocity fields in radiation calculations, and the effects of changes in operating conditions.

The second example involves a prediction of the effects of concentration fluctuations in a free turbulent propane/air diffusion flame. Becker and Brown [74] have mapped the fields of mean impact pressure $\bar{\rho}\bar{U}_x^2$ for an air/air jet and for a propane/air flame jet, all on the same jet nozzle, and for the flame they also obtained the mean temperature field $\bar{T}$. These results, similar data of other authors, and theoretical requirements of the conservation of momentum, energy, and material indicate that the following approximations are reasonable:

1. The fields of the flux densities of momentum and nozzle-gas material, namely $\bar{\rho}\bar{U}_x^2$ and $\bar{\rho}\bar{U}_x\bar{W}_1$, are the same in the flame jet as in the cold system.

2. The field of $\bar{\rho}\bar{U}_x\bar{W}_1$ is the same as that of $\bar{\rho}\bar{U}_x\bar{H}$ in the flame jet, where $\bar{H} \equiv (\bar{h} - h_{2,o})/(h_{1,o} - h_{2,o})$.

Becker, Hottel and Williams [21] have mapped the fields of $\bar{Y}_1$ and $\overline{y_1^2}$ in the cold jet. It is further supposed that:

3. The field of $\hat{w}_1/\bar{W}_1$ in the flame jet is the same as that of $\hat{y}_1/\bar{Y}_1$ in the cold jet.

Finally, it is supposed that:

4. The mass fraction W_1 of nozzle-gas material (fuel atoms) in the flame is normally distributed about its mean, so

$$f(W_1) = \frac{1}{2\sqrt{\pi}\hat{w}_1} \exp(-\overline{w_1^2}/2\hat{w}_1^2) \; .$$

5. The flame jet system is locally in thermodynamic equilibrium. Then $W_i = W_i(W_1)$, or $X_i = X_i(W_1)$.

It is possible, under these assumptions, to compute the intensity field $\hat{\theta}/(\bar{T} - T_{2,o})$ for temperature and the intensity fields $\hat{x}_i/\bar{X}_i$ for the mole fractions of various molecular components. The calculations have been done by Becker [49] and give as reasonable a prediction of the effects of turbulence as can presently be obtained. The assumption that all species are locally in equilibrium is definitely more satisfactory than the one in the first example above – that fuel and oxygen are fully reacted to the extent that oxygen is available. This approach can be incorporated into the full mathematical modelling of diffusion flames. One would accordingly use the transport equations for the mass fractions of feed materials, W_n, and the equation for total enthalpy h; the fields of temperature and molecular species mole fractions would at the same time be computed from the values of W_n and h.

Similar computations on a somewhat simpler model have been presented by Bilger and Kent [74] for a hydrogen/air diffusion flame. They supposed that (i) $\hat{w}_1/\overline{W}_1 = 0.5$, and (ii) the hydrogen is everywhere fully reacted to the extent that oxygen is available.

Becker limited his computations to the core region of low intermittency, $\bar{\delta} \approx 1$. Grandmaison's measurements [54] of skewness $\overline{y_1{}^3}/\hat{y}_1{}^3$ and kurtosis $\overline{y_1{}^4}/\hat{y}_1{}^4$ in a cold jet indicate that the assumption of a Gaussian p.d.f. is quite reasonable for this region. However, Becker's model can presumably be improved by introducing the measured p.d.f.'s for cold systems. A simpler, and probably adequate approach is to place spikes (impulse functions) at the physical limits of the p.d.f. where $W_1 = 0$ or $W_1 = 1$, and thus satisfy

$$\int_0^1 f(W_1)dW_1 = 1 \ ;$$

this is essentially the idea used by Spalding [71] in computing the concentration fluctuation field of a cold jet.

7. INVESTIGATIONS WITH MARKER NEPHELOMETRY

7.1 General

A brief review will now be given to indicate all the systems that have been studied by marker nephelometry and the measurements that were made. A few remarks will also be made on systems not yet investigated (particularly turbulent diffusion flames) and measurements not yet attempted.

The only investigations that have been made of one-feed and three-feed problems have already been described: the premixed flame of Williams *et al.* [57] in Sec.5.2, and the system of two interacting free air jets of Becker and Booth [60] in Sec.5.4. The other investigations have all been of constant-density two-feed systems. It will be convenient, in discussing these, to revert to the notation of Section 4 and let Γ be the concentration of marked-stream material, with $Y = \Gamma/\Gamma_o = \Gamma/\rho_o$. Subscripts to identify feeds are redundant here, since in non-dilational two-feed mixing only one feed needs to be marked to determine the concentration fields of both feed materials.

7.2 The Free Non-swirling Jet

Rosensweig *et al.* [19], as noted earlier, carried out the pioneering investigation of marker nephelometry and its applications. The feasibility was demonstrated of determining the mean concentration $\overline{\Gamma}$, the concentration fluctuation intensity $\hat{\gamma}$, the spectral density function $E(\kappa)$, the two-point spatial correlation function $R(\mathbf{x}_A; \mathbf{x}_B)$, and the integral length scale Λ_γ. The measurements were made in a round free jet of air mixing with air at Reynolds numbers from 16,000 to 57,000.

Becker *et al.* [20,21] refined the technique and carried out a further study of the round free jet with the object of characterizing the behaviour of this flow structure as accurately as possible. The fields of $\overline{\Gamma}$, $\hat{\gamma}$, and the intermittency factor $\bar{\delta}$ were mapped, and various measurements of $E(\kappa)$ and $R(\mathbf{x}_A;\mathbf{x}_B)$ were made. The jet nozzle diameter in these studies was 0.635 cm, and the Reynolds number was 54,000, high enough for viscous influences on the mean flow and energy-containing parts of the turbulence to be negligible. Recently Grandmaison [54] has repeated these measurements on a very much larger jet: the nozzle diameter was 7.14 cm, measurements were carried up to 3 m downstream, and the Reynolds number was 240,000. Very close agreement was found with Becker *et al.*. Grandmaison also measured the fields of the skewness $\overline{\gamma^3}/\hat{\gamma}^3$ and the kurtosis $\overline{\gamma^4}/\hat{\gamma}^4$, as noted in Sec.4.4, and he extended the measurements of $E(\kappa)$ up to $\kappa\Lambda_\gamma = 10^4$ without finding evidence of Batchelor's -1 power law for the viscous-convective subregime. The last result was a surprise, and further work is being done to check it. Measurements of the probability density function of γ are also in progress.

Whitelaw and Melling [23] are doing work in which they are combining marker nephelometry with laser-doppler velocimetry to measure components of the concentration-velocity covariance $\overline{\gamma\mathbf{u}}$.

7.3 *The Free Swirling Jet*

Grandmaison [54], working in the author's laboratory, has studied swirling free jets of air at swirl numbers ranging from 0 to 1.4. The fields of $\overline{\Gamma}$, $\hat{\gamma}$, and $\bar{\delta}$ were mapped, and $E(\kappa)$, $R(\mathbf{x}_A;\mathbf{x}_B)$ and Λ_γ were measured at critical locations. Concentration fluctuation levels $\hat{\gamma}/\overline{\Gamma}$ on the jet centreline were found to be similar to those in non-swirling jets. The swirling jets spread more rapidly, and show a greater region of intermittency.

7.4 *Dust-laden Non-swirling Free Jets*

Raichura [44] and Rajani [45], working with Langer*, have studied a dust-laden free jet of air mixing with air. The dust loading was varied between 0 and 2.2 kg/kg. Nephelometry was used to map both the dust concentration field and the concentration field of oil condensation smoke introduced to mark the nozzle air. Measurements were made of mean concentration, concentration fluctuation intensity, and spectrum. Two complications arise in the determination of fluctuations in this experiment: (i) the "marker" shot noise of the dust may be very high, and (ii) the correlation between dust and smoke concentration fluctuations is significant and requires accounting for. Rajani also measured the mean velocity fields of the smoke and dust particles by means of the laser-doppler technique.

7.5 *The Non-swirling Axisymmetrical Ducted Jet*

Becker *et al.* [52,55,72,76] have studied the fields of $\overline{\Gamma}$, $\hat{\gamma}$ and $\bar{\delta}$ in an air/air ducted jet system under conditions ranging from nearly total recirculation to just beyond the vanishing point of the recirculation phenomenon. The study was conceived as a cold modelling of the cement-kiln type of tunnel furnace, and

* Dr. G. Langer, Department of Mechanical Engineering, Queen Mary College, University of London, Mile End Road, London E1 4NS.

the conditions were those of greatest interest in this context. Measurements were also made of $E(\kappa)$, Λ_γ and of $R(\mathbf{x}_A; \mathbf{x}_B)$.

Catalano [22] and Shaughnessy [24], working with Morton*, have done similar work for some conditions of no recirculation. Catalano also succeeded in obtaining measurements of the concentration-velocity covariance $\overline{\gamma u_x}$ in the near field of the jet by combining marker nephelometry with laser-Doppler velocimetry. Values of the correlation function $R_{\gamma x} \equiv \overline{\gamma u_x}/\hat{\gamma}\hat{u}_x$ ranged from -0.1 to 0.5.

7.6 *Jets in a Transverse Pipe Flow*

Rathgeber [77], working in the author's laboratory, has mapped the fields of $\overline{\Gamma}$, $\hat{\gamma}$, and $\overline{\delta}$, and made selected measurements of $E(\kappa)$, for an air/air system consisting of a jet injected from a tube through a hole in the pipe wall into a transverse pipe flow. The inside diameter ratio of the injector tube and the pipe, and the mass-flux ratios of the two streams, were varied over a broad range. This system has the interesting feature that the bent-over jet is modified by the primary flow to form a pair of counter-rotating line vortices.

Work by K.I.M. Smith on two opposed jets is also in progress; he is studying the field of $\overline{\gamma_1 \gamma_2}$ by marking first one jet, then the other, and finally both.

7.7 *Turbulent Diffusion*

Becker *et al.* [51] studied turbulent diffusion from a point source on the axis of a pipe flow. The fields of $\overline{\Gamma}$ and $\hat{\gamma}$ were mapped and some spectra $E(\kappa)$ were obtained. Values of velocity fluctuation intensity and Lagrangian integral length scale were computed from the field of $\overline{\Gamma}$, as mentioned earlier in Sec.4.4. A novel feature of the measurements was that a cross-section of the diffusion plume was sheet-illuminated and the quantity

$$\Omega \equiv \int_S \Gamma \, dS$$

detected. Values of $\overline{\Omega}$ and $\hat{\omega}$ were determined, and frequency spectra of the fluctuations $\omega \equiv \Omega - \overline{\Omega}$ were measured.

Yang and Meroney† [46] have performed an interesting experiment in which a smoke-filled gas bubble was introduced through a column of water to burst at the floor of a wind tunnel. Time-dependent concentrations were measured at points downstream. The smoke ideally entered as from a point source, into a thick turbulent boundary layer. They used a portable, very compact nephelometer probe of novel design, described in another publication [47].

Liu and Karaki** [48] measured $\overline{\Gamma}$ and $\hat{\gamma}$ downstream from an elevated point source in a windstream over a water surface ruffled by wind-generated waves. They also developed a very compact nephelometer probe of novel design.

---

* Dr. J.B. Morton, Department of Engineering Science and Systems, University of Virginia, Charlottesville, Virginia 22901.

† Dr. R.N. Meroney, College of Engineering, Colorado State University, Fort Collins, Colorado 80521.

** Dr. S. Karaki, Department of Civil Engineering, Colorado State University, Fort Collins, Colorado 80521.

This probe intrudes mechanically into the measurement region, and is thus subject to flow interference errors; on the other hand, it appears to be exceptionally well suited for situations where signal modulation due to light extinction by the field along the optical path is a problem.

7.8 *Turbulent Diffusion Flames*†

The results of Williams, Hottel and Gurnitz [57] on a premixed flame indicate that marking with magnesium oxide particles is satisfactory for work on flames at ordinary combustion temperatures. In the premixed flame $\Gamma_b/\Gamma_u = \rho_b/\rho_u$, where b refers to the burnt gas and u to the unburnt, and the value of ρ_b/ρ_u can be checked by other means. The values obtained for ρ_b/ρ_u were reasonably good. Aluminium oxide particles should also be acceptable.

Work has been in progress in the author's laboratory for several years towards the study of a free-jet turbulent diffusion flame by marker nephelometry. Apparatus, including a MgO smoke generator, was developed with R. Keshavan. Preliminary measurements on a propane/air flame indicated the feasibility of the approach but also the need for a fairly powerful laser. A 600 mW argon ion laser was acquired, the work was continued by S. Yamazaki, and a thorough study of soot concentration fields in propane/air flames has just been completed. However, the soot was found to constitute too thick a background to allow unambiguous measurements of the MgO marker concentration field. It is now planned to switch from hydrocarbon fuel gases to either hydrogen or a mixture of carbon monoxide and a trace of hydrogen, in order to have an essentially soot-free flame. The experiment should then succeed; the performance of the MgO smoke marker and of the system as a whole was reasonably good apart from the soot problem.

The interpretation of results from such a study will not be straightforward, since the mass concentration of jet-feed material (fuel atoms) is strongly influenced by both dilation (due to temperature variation) and mixing, as shown by (5.1.1). However, the fields of $\overline{\Gamma}$ and $\hat{\gamma}$ do have a basic significance in relation to theory; they constitute something that can be predicted by mathematical models together with other quantities of interest, and hence the results can be used to calibrate and check such models, and perhaps help guide their formulation. It will also be interesting to see the behaviour of the pdf of γ and the field of the intermittency factor $\overline{\delta}$.

The experiment should, to provide most information, be carried out with all possible feed markings – first the fuel gas, then the air, and finally both. With subscript 1 denoting the fuel and 2 the air, (5.1.1) gives:

$$Y_1 = (M/M_{1,0})(T_{1,0}/T)W_1 \tag{7.8.1}$$

$$Y_2 = (M/M_{2,0})(T_{2,0}/T)W_2 \; . \tag{7.8.2}$$

With uniform marking of both feeds (i.e. $\Gamma^*_{1,o} = \Gamma^*_{2,o}$)

$$Y \equiv Y_1 + Y_2 = \frac{M}{M_{1,0}} \frac{T_{1,0}}{T} \left\{ W_1 + W_2 \frac{M_{1,0}}{M_{2,0}} \frac{T_{2,0}}{T_{1,0}} \right\} . \tag{7.8.3}$$

† Note added in proof. Marker nephelometry has been successfully applied to turbulent diffusion flames in a recent study at the Engler-Bunte-Institute, University of Karlsruhe, Combustion

Let $T_{1,o} = T_{2,o} \equiv T_o$. In the case of CO burning with air, $M_{1,o} \approx M_{2,o}$. Then, to a good approximation,

$$Y_1 = (M/M_o)(T_o/T)W_1 \tag{7.8.4}$$

$$Y_2 = (M/M_o)(T_1/T)W_2 \tag{7.8.5}$$

$$Y = (M/M_o)T_o/T \ , \tag{7.8.6}$$

where, say, $M_o \equiv M_{2,o}$. The statistical characterization of the fields can be done as follows. In the case of Y_1, write

$$Y_1 \approx \frac{\overline{M}'}{M_o}\frac{T_o}{\overline{T}'}\,\overline{W}_1'\left\{1-\frac{\theta'}{\overline{T}'}\right\}\left\{1+\frac{w_1'}{\overline{W}_1'}\right\}, \qquad Y_1 > 0 \ , \tag{7.8.7}$$

where $Q = W_1$, T, or M and

$$\overline{Q}' \equiv \lim_{\tau\to\infty}\frac{1}{\overline{\delta}\tau}\int_t^{t+\tau}\delta Q dt \ , \tag{7.8.8}$$

The quantity δ is the intermittency function defined in Sec.4.4, here

$$\delta = (1, Y_1 > 0;\ 0, Y_1 = 0) \ ; \tag{7.8.9}$$

$$\overline{\delta}\overline{W}_1' = \overline{W}_1, \quad \overline{\delta}\,\overline{T}' = \overline{T}, \quad \overline{\delta}\,\overline{M}' = \overline{M} \ , \tag{7.8.10}$$

$$w_1' \equiv W_1 - \overline{W}_1', \quad \theta' \equiv T - \overline{T}' \ . \tag{7.8.11}$$

Molecular weight fluctuations are neglected. Then

$$\overline{Y}_1' = \frac{\overline{M}'}{M_o}\frac{T_o}{\overline{T}'}\,\overline{W}_1'\left\{1-\frac{\overline{\theta' w_1'}}{\overline{T}'\overline{W}_1'}\right\} \ , \tag{7.8.12}$$

$$\overline{y_1'^2} = \left\{\frac{\overline{M}'}{M_o}\frac{T_o}{\overline{T}'}\,\overline{W}_1'\right\}^2\left\{\frac{\overline{\theta'^2}}{\overline{T}'^2}+\frac{\overline{w_1'^2}}{\overline{W}_1'^2}-2\frac{\overline{\theta' w_1'}}{\overline{T}'\overline{W}_1'}+\frac{\overline{\theta'^2 w_1'}}{\overline{T}'^2\overline{W}_1'} - \frac{\overline{\theta' w_1'^2}}{\overline{T}'\overline{W}_1^2}+\frac{\overline{\theta'^2 w_1'^2}}{\overline{T}'^2\overline{W}_1'^2}\right\} \tag{7.8.13}$$

where $\overline{\delta}\,\overline{Y}_1' = \overline{Y}_1$. Analogous expressions apply to $\overline{Y}_2'$ and $\overline{y_2'^2}$. Finally, from Y,

$$\overline{Y}' = \frac{\overline{M}'}{M_o}\frac{T_o}{\overline{T}'} \ , \tag{7.8.14}$$

---

† *(cont.)*
Technology Department, Director Prof. Dr.-Ing. R. Guenther, Karlsruhe, Germany: Ebrahimi, I. and Kleine, R., The nozzle fluid concentration field in round free jets and jet diffusion flames, Sixteenth Symposium (International) on Combustion, The Combustion Institute, 1977.

$$\overline{y'^2} = \left\{\frac{\overline{M}'}{M_o}\frac{T_o}{\overline{T}'}\right\}^2\left\{\frac{\overline{\theta'^2}}{\overline{T}'^2}\right\} . \tag{7.8.15}$$

The measurements indicated require conditional sampling. The properties defined in these relations are precisely those of greatest interest – the properties of the flame-jet fluid – the material inside the turbulent jet boundary.

The temperature-composition double covariance can be represented as

$$\overline{\theta' w_1'} = R_{\theta 1}\hat{\theta}_1'\hat{w}_1' .$$

The value of $R_{\theta 1}$ may approach $+1$ or -1; it is expected to be negative in the region of super-stoichiometric $\overline{W}_1'$, positive in the sub-stoichiometric region, and around zero in the region of the stoichiometric surface. The two triple covariances in (7.8.13) can similarly be represented in terms of $\overline{\theta' w_1'^2}/\hat{\theta}'\hat{w}_1'^2$ and $\overline{\theta'^2 w_1'}/\hat{\theta}'^2\hat{w}_1$; these quantities should behave rather like the skewnesses $\overline{\theta_1'^3}/\hat{\theta}_1'^3$ and $\overline{w_1'^3}/\hat{w}_1'^3$, and can be expected to be negligible. The quadruple covariance $\overline{\theta'^2 w_1'^2}$ can be represented in terms of $\overline{\theta'^2 w_1'^2}/\hat{\theta}'^2\hat{w}_1'^2$ which should behave quite like the kurtoses of θ' and w_1'; it can therefore be expected that roughly

$$\overline{\theta'^2 w_1'^2} = 3\hat{\theta}'^2\hat{w}_1'^2 .$$

Let us suppose that $\hat{\theta}'/\overline{T}' = \hat{w}_1'/\overline{W}_1' = 0.25$ – a reasonable guess at the maximum values, according to the calculations of Becker [49]. The term in $\overline{\theta' w_1'}$ is then less than 6% of $\overline{Y}_1'$ in (7.8.12) and can be considered negligible. In (7.8.13), however, it is probably negligible only in the neighbourhood of the stoichiometric surface of $\overline{W}_1'$. The triple covariances in (7.8.13) can reasonably be neglected. The term in $\overline{\theta'^2 w_1'^2}$ should be less than 25% of $\overline{y_1'^2}$, but probably never negligible.

There are also available, however, the equations for $\overline{Y}_2'$ and $\overline{y_2'^2}$ and the relations

$$\overline{W}_1' + \overline{W}_2' = 1 ; \qquad w_1' + w_2' = 0 .$$

Thus

$$\overline{Y}_2' = \frac{\overline{M}'}{M_o}\frac{T_o}{\overline{T}'}\overline{W}_2'\left\{1 + \frac{\overline{\theta' w_1'}}{\overline{T}'\overline{W}_2'}\right\} \tag{7.8.16}$$

and

$$\overline{y_2'^2} = \left\{\frac{\overline{M}'}{M_o}\frac{T_o}{\overline{T}'}\overline{W}_2'\right\}^2\left\{\frac{\overline{\theta'^2}}{\overline{T}'^2} + \frac{\overline{w_1'^2}}{\overline{W}_2'^2} + 2\frac{\overline{\theta' w_1'}}{\overline{T}'\overline{W}_2'} - \frac{\overline{\theta'^2 w_1'}}{\overline{T}'^2\overline{W}_2'} - \frac{\overline{\theta' w_1'^2}}{\overline{T}'\overline{W}_2'^2} + \frac{\overline{\theta'^2 w_1'^2}}{\overline{T}'^2\overline{W}_2'^2}\right\} . \tag{7.8.17}$$

The value of $\overline{M}'$ can be estimated as the thermodynamic equilibrium value $\overline{M}' = \overline{M}'(\overline{W}_1', \overline{T}')$. It should then be possible, using the above system of equations, to determine $\overline{W}_1'$, $\overline{W}_2'$, $\overline{T}'$, and $\hat{\theta}'$ with reasonable accuracy, and to obtain moderately good estimates of $\hat{w}_1' = \hat{w}_2'$.

Because of the necessity of marking the air, as well as the fuel, the experiment described would have to be done on a confined flame-jet system.

7.9 The Technique of Wilhelm and Coworkers

A form of marker nephelometry very different from that described here has been developed by Wilhelm† and coworkers [77–79]. Their quite remarkable technique is applicable to liquid/liquid mixing, and the marker is itself a liquid. Christiansen [79], for example, studied a water/water system in which one water feed was marked with ethylene glycol. The technique involves analysis of the forward light scattering, and has a problem of resolution in that the information is all contained in a very small range of scattering angles. The result is an estimate of the three-dimensional γ^* wave-number spectrum from which $\overline{\gamma^{*2}}$ can be computed. The system investigated by Christiansen was a mixing analogue of the classical experiments on the decay of grid turbulence; he bled marked water into a turbulent pipe flow from an array of hypodermic injectors and studied the decay of the marker concentration fluctuations.

7.10 Laser-Raman Nephelometry

One of the newest areas of development in the optical detection of scalar fields is laser-Raman nephelometry. This offers the exciting prospect of directly detecting temperature and the concentrations of many molecular species. However, extremely intense irradiation is required, and it remains to be proven that radiation interference effects, particularly electrical breakdown, are then unimportant. It also remains to be seen whether adequate signal/noise ratios can be attained for turbulence measurements. The author would be interested in working with this technique if the capital costs were not so high; he hopes, at any rate, that it will be brought to success by others, for the detail of information it can potentially provide is considerable. A book on the laser-Raman technique has been published [18] and Goulard [80] has written on its applications. Hartley [81,82] has made concentration measurements in transient gas mixing, and Setchell [83] has obtained results on mean concentration and temperature in flames.

7.11 Comparison with Results Obtained by Other Techniques

A critical comparison between results obtained by marker nephelometry and those found by other techniques is mainly important with regard to turbulent fluctuations. There is no significant disagreement regarding mean concentration fields, as indicated by Becker, Hottel and Williams' review of data on free jets [21], nor would any be expected.

There have only been a few studies reported on fluctuations in scalar fields with which direct comparisons are possible. It is expected that in gas mixing the fields of temperature or concentration resulting from the marking of a given feed with thermal energy ("heat") or gaseous material should be virtually the same if the density fields are reasonably similar. There has been a number of studies of

† Professor R.H. Wilhelm (deceased), Department of Chemical Engineering, Princeton University.

axisymmetrical jets in which temperature fluctuations were measured by hot-wire thermometry. The data of Corrsin and Uberoi [84] on $\hat{\theta}/(\bar{T}-T_\infty)$ in a free air/air jet, Table I, agree with those of Becker *et al.* [21] on $\hat{\gamma}/\bar{\Gamma}$ at small x/D,

Table I. Intensity of concentration or temperature fluctuations on the centrelines of axisymmetrical free jets (f.j.) and ducted jets (d.j.). Key to references: BHW = Becker, Hottel and Williams, WL = Way and Libby, M = McQuaid, CU = Corrsin and Uberoi, WD = Wilson and Danckwerts, AB = Antonia and Bilger.

	$\hat{\gamma}/\bar{\Gamma}$			$\hat{\theta}/(\bar{T}-T_\infty)$		
x/D	BHW [21] f.j.	WL,M [90,91] f.j.	BHW [52] d.j.	CU [86] f.j.	WD [87] f.j.	AB [89] d.j.
10	.170	.36	.150	.175	.11	
15	.177	.34	.164	.173	.14	
20	.185	.36	.172	.165	.15	
30	.203	—	.183	.140	.17	.20
40	.208	—	.197	—	.175	.21
60	—	—	.215	—	.18	.22
∞*	.222	—	—	—	.18	—

* Extrapolated values.

but then diminish while Becker *et al.*'s rise. Wilson and Danckwerts [85] found values of $\hat{\theta}/(\bar{T}-T_\infty)$ around 0.03 below Becker *et al.*'s $\hat{\gamma}/\bar{\Gamma}$ at x/D between 10 and 40, and smaller by 0.04 in the limit $x \to \infty$. Similar results for confined jets have been obtained by Antonia and Bilger [89] for $\hat{\theta}/(\bar{T}-T_\infty)$ and Becker *et al.* [52] for $\hat{\gamma}/\bar{\Gamma}$. Table I shows reasonable agreement for cases with similar operating conditions. To some extent the differences seen here in the results of investigators using hot-wire thermometry probably reflect effects of improvements in the technique, but in the near and middle regions of the jets they may also be due to differences in the nozzle air temperature: the initial jet temperature as 170°C for the experiment of Corrsin and Uberoi, 225°C for that of Wilson and Danckwerts but only 34°C for that of Antonia and Bilger. From the latter viewpoint, the results of Antonia and Bilger are the most directly comparable with Becker *et al.*.

Substantial disagreement is found with the results obtained by Way and Libby [87] and McQuaid [88] who used hot wires – similar to those employed in hot-wire thermometry – to measure concentration fluctuations in a round free jet. Their values of $\hat{\gamma}/\bar{\Gamma}$ on the centreline, Table I, are proportional to those of Becker *et al.*, but twice as high. There is no readily apparent explanation for this discrepancy.

Data on free jets at high Reynolds number have been obtained by marker nephelometry in three different studies now [21,54,60]. Grandmaison's results

[54] for a jet on a 7 cm dia flow nozzle at a Reynolds number of 220,000 show excellent agreement with those of Becker *et al.* [21], as mentioned earlier. The data of Becker and Booth [60] for a jet from a long tube of 0.63 cm i.d. at a Reynolds number of 70,000 agree closely with Becker *et al.* and Grandmaison for $x/D > 25$, i.e. far enough downstream for the different initial conditions of the experiments to be uninfluential.

On the whole, the accumulated evidence from all the studies mentioned above seems strongly to suggest that marker nephelometry on gas mixing processes indeed performs as theory predicts that it should. No results have yet been found that throw doubt upon the validity of the technique. On the contrary, it appears that its validity has been well established, and results obtained with it by careful measurements can be regarded as accurate within uncertainty limits that can be estimated from theory and experiment.

8. CONCLUDING REMARKS

The article was originally conceived as one that would describe, in general terms, the writer's studies in mixing, turbulence and combustion. Only in assembling the material did the central role played by marker nephelometry fully emerge. Not only had the technique been the experimental basis of much of the work; it had significantly shaped the direction of many of the investigations and the ideas that developed from them. It soon appeared that a thorough presentation of the theory and application of marker nephelometry itself would have greater novelty and usefulness than the review originally contemplated. It has been seen herein that the technique puts emphasis on the mass concentration fields of feed materials. This emphasis particularly fits the investigation of gas-phase processes of mixing and dilation, including those in turbulent diffusion flames and similar chemical reaction systems where reaction rates tend to be diffusion-controlled.

The applications discussed in the article have shown that marker nephelometry is a versatile tool. The long catalogue of sources of error provided in Section 3 should not deter the prospective user; there is usually an ample middle ground where good results are easily obtained. Moreover, all techniques of turbulence investigation have numerous pitfalls for the unwary.

There are, though, viable alternatives that should be considered. The technique of hot-wire temperature detection – hot-wire thermometry – for instance, is well developed and allows the investigation of gas/gas mixing using thermal energy ("heat") as a marker. This has the advantage that the scalar dissipation spectrum is accessible. The work of Antonia and coworkers [86,89,90] illustrates what can currently be accomplished. The equipment costs for marker nephelometry and hot-wire thermometry are roughly the same, and the instrumentation for signal processing is largely interchangeable. The reasons for preferring marker nephelometry for some situations are often the same as those for preferring laser velocimetry to hot-wire velocimetry: e.g., non-interference with the flow, applicability in difficult environments, minimal need for calibration, and naturally excellent linearity and frequency response.

Some other techniques for the detection of scalar fields were mentioned in the Introduction. Electrical conductivity probes appear to be the most sensitive tool available for water systems, as shown by the work of Gibson and Schwartz [8] and of Tsujikawa and coworkers [9,10].

A continuing goal in the author's work has been to obtain a better understanding of the structure of turbulent diffusion flames. This motivation has also led to the development of other experimental techniques that are applicable to flames. An analysis of the response of pitot probes in turbulent streams [50,74] suggested a differential pitot probe for the detection of velocity fluctuations. Similar work is in progress on static pressure probes.

The theory of mixing and chemical reaction in gaseous turbulent fields under the assumptions of equal binary diffusivities and diffusion-limited reaction rates is now reasonably well advanced. Early work on this problem was done by Hawthorne, Weddell and Hottel [59], Richardson, Howard and Smith [91], and Toor [92]. The recent work by Bilger and Kent [75] and the author [49] using Gaussian pdf's for the key concentration or composition fluctuations has been mentioned. In [93] Professor Libby discusses other models of the pdf, a topic he takes up also in his article in this volume. The next stage of development should be to introduce some chemical reactions whose time scales are comparable with those of the turbulence. Work in this area has already been done by Corrsin [94], Toor and coworkers [95–97], Yieh [98], Patterson [99], Borghi [100, 101], and Bilger [63]. The effect of unequal diffusivities might be explored in a few trial cases, but it is doubtful that the improvement would be practically significant for most turbulent systems. The general problem of multicomponent mass transfer in turbulent flows has been examined by Stewart and Prober [102], Von Behren, Jones and Wasan [103], and Stewart [104].

REFERENCES

1. Fristrom, R.M. and Westenberg, A.A. "Flame Structure", McGraw-Hill, 1965.
2. Leeper, C. Sc. D. Thesis, Massachusetts Institute of Technology, 1954.
3. Tiné, G. "Gas Sampling and Chemical Analysis in Combustion Processes", Pergamon, 1961.
4. Beér, J.M. and Chigier, N.A. "Combustion Aerodynamics", Halstead Press, 1972.
5. Chedaille, J. and Brand, Y. "Industrial Flames, Vol.1: Measurements in Flames", Edward Arnold, 1972.
6. Bilger, R.W."Probe Measurements in Turbulent Combustion", Rept. No.PURDU-CL-75-02, Combustion Laboratory, School of Mechanical Engineering, Purdue University, West Lafayette, Indiana, 1975.
7. Lamb, D.E., Manning, F.S. and Wilhelm, R.H. "Measurement of Concentration Fluctuations with an Electrical Conductivity Probe", *A.I.Ch.E.J.* 6, 682-685, 1960.
8. Gibson, C.H. and Schwartz, W.H. "The Universal Equilibrium Spectra of Turbulent Velocity and Scalar Fields", *J. Fluid Mech.* 16, 365-385, 1963.
9. Tsujikawa, H. and Uraguchi, Y. "The Discrepancy of Concentration Measured by the Electrical Conductivity Method Compared to that of the Volumetric Mean in a Fluctuating Concentration Field", *J. Chem. Eng. Japan* 6, 92-96, 1973.
10. Tsujikawa, H., Mishima, S., Nagamoto, H., Miyawaki, O. and Uraguchi, Y. "Direct Measuring Method of Concentration Spectra by Electrical Conductivity", *J. Chem. Eng. Japan* 7, 299-303, 1974.
11. Lockwood, F.C. and Odidi, A.O. "Measurement of Mean and Fluctuating Temperature and of Ion Concentration in Round Free-jet Turbulent Diffusion and Premixed Flames", *Fifteenth Symposium (International) on Combustion*, pp.561-571, The Combustion Institute, 1975.

12. Page, F.M., Roberts, W.G. and Williams, H. "An Experimental Study of the Interaction of Chemical Kinetic Effects and Turbulent Flow in Flames", *Fifteenth Symposium (International) on Combustion*, pp.617-624, The Combustion Institute, 1975.
13. Liu, H. "The Streaming Potential Fluctuations in a Turbulent Pipe Flow", *A.I.Ch.E.J.* 13, 644-649, 1967.
14. Lee, J. and Brodkey, R.S. "Light Probe for the Measurement of Turbulent Concentration Fluctuations", *Rev. Sci. Instrum.* 34, 1086-1090, 1963.
15. Lee, J. and Brodkey, R.S. "Turbulent Motion and Mixing in a Pipe", *A.I.Ch.E.J.* 10, 187-193, 1964.
16. Guenther, R. and Simon, H. "Turbulence Intensity, Spectral Density Functions and Eulerian Scales of Emission in Turbulent Diffusion Flames", *Twelfth Symposium (International) on Combustion*, pp.1069-1079, The Combustion Institute, 1969.
17. Kuhnert, D. and Günther, R. "Turbulent Scale of OH Radiation", *Combustion Institute European Symposium 1973*, pp.518-523, Academic Press, 1973.
18. Lapp, M. and Penney, C.M. "Laser Raman Gas Diagnostics", Plenum Press, 1973.
19. Rosensweig, R.E., Hottel, H.C. and Williams, G.C. "Smoke-scattered Light Measurements of Turbulent Concentration Fluctuations", *Chem. Eng. Sci.* 15, 111-129, 1961.
20. Becker, H.A., Hottel, H.C. and Williams, G.C. "On the Light-scatter Technique for the Study of Turbulence and Mixing", *J. Fluid Mech.* 30, 259-284, 1967.
21. Becker, H.A., Hottel, H.C. and Williams, G.C. "The Nozzle-fluid Concentration Field of the Round, Turbulent, Free Jet", *J. Fluid Mech.* 30, 285-303, 1967.
22. Catalano, G.D. "An Experimental Investigation of an Axisymmetric Jet in a Coflowing Stream", M.Sc. Thesis, University of Virginia, 1975.
23. Whitelaw, J.H., Private communication, 1975.
24. Shaughnessy, E.J. "Measurement of Particle Diffusion in a Turbulent Jet by Laser Light Scattering", Ph.D. Thesis, University of Virginia, 1975.
25. Hottel, H.C. and Sarofim, A.F. "Radiative Transfer", McGraw-Hill, 1967.
26. Siegel, R. and Howell, J.R. "Thermal Radiation Transfer", McGraw-Hill, 1972.
27. Van de Hulst, H.C. "Light Scattering by Small Particles", Wiley, 1957.
28. Kirker, M. "The Scattering of Light and Other Electromagnetic Radiation", Academic Press, 1969.
29. Plass, P.N. "Mie Scattering and Absorption Cross-sections for Absorbing Particles", *Appl. Optics* 5, 279-285, 1966.
30. Blevin, W.R. and Brown, W.J. "Effect of Particle Separation on The Reflectance of Semi-infinite Diffusers", *J. Opt. Soc. Amer.* 51, 129-134, 1961.
31. Churchill, S.W., Clark, G.C. and Sliepcevich, C.M. "Light-scattering by Very Dense Mono-dispersions of Latex Particles", *Discussions of the Faraday Soc.* 30, 192-199, 1960.
32. Parzen, E. "Stochastic Processes", Holden-Day, 1962.
33. Philips Electron Tube Division "Philips Photomultiplier Tubes", The Philips Co., Holland, 1964.
34. Sharpe, J. and Stanley, V.A. "An Introduction to the Photomultiplier", EMI Electronics Ltd., 1974.
35. Saffman, P.G. and Turner, J.S. "On the Collision of Drops in Turbulent Clouds", *J. Fluid Mech.* 1, 16,
36. Howarth, W.J. "Coalescence of Drops in a Turbulent Flow Field", *Chem. Eng. Sci.* 19, 33-38, 1964.
37. Hinze, J.O. "Turbulent Fluid and Particle Interaction", WTHD 32, Laboratorium Voor Aero-en-Hydrodynamica, Delft University of Technology, 1971.
38. Birkhoff, G. "Hydrodynamics", Dover, 1955.
39. Batchelor, G.K. "An Introduction to Fluid Dynamics", Cambridge University Press, 1967.
40. Whytlaw-Gray, R. and Patterson, H.S. "Smoke", Edward Arnold, 1932.
41. Patankar, S.V. and Spalding, D.B. "Heat and Mass Transfer in Boundary Layers", Morgan-Grampian, 1967.
42. Launder, B.E. and Spalding, D.B. "Mathematical Models of Turbulence", Academic Press, 1972.
43. Gosman, A.D., Pun, W.M., Runchal, A.K., Spalding, D.B. and Wolfshtein, M. "Heat and Mass Transfer in Recirculating Flows", Academic Press, 1969.
44. Raichura, R.C. "Measurement of Turbulent Mixing in a Free Dust-laden Air Jet", Ph.D. Thesis, University of London, 1968.

45. Rajani, J.B. "Turbulent Mixing in a Free Air Jet Carrying Solid Particles", Ph.D. Thesis, University of London, 1972.
46. Yang, B.T. and Meroney, R.N. "On Diffusion from an Instantaneous Point Source in a Neutrally Stratified Boundary Layer with a Laser Light Scattering Probe", Project Themis Tech. Rept. No.20, Fluid Dynamics and Diffusion Laboratory, Engineering Research Center, College of Engineering, Colorado State University, Fort Collins, Colorado, 1972.
47. Yang, R.T. and Meroney, R.N. "A Portable Laser Light-scattering Probe for Turbulent Diffusion Studies", *Rev. Sci. Instrum.* **45**, 210-215, 1974.
48. Liu, H.T. and Karaki, S. "An Optical System for the Measurement of Mean and Fluctuating Concentrations in an Airstream", *J. Phys. E: Sci. Instruments* 5, 1165-1168, 1972.
49. Becker, H.A. "Effects of Concentration Fluctuations in Turbulent Diffusion Flames", *Fifteenth Symposium (International) on Combustion*, pp.601-615, 1975.
50. Becker, H.A. and Brown, A.P.G. "Response of Pitot Probes in Turbulent Streams", *J. Fluid Mech.* **62**, 85-114, 1974.
51. Becker, H.A., Rosensweig, R.E. and Gwozdz, J.R. "Turbulent Dispersion in a Pipe Flow", *A.I.Ch.E.J.* **12**, 964-972, 1966.
52. Becker, H.A., Hottel, H.C. and Williams, G.C. "Concentration Fluctuations in Ducted Turbulent Jets", *Eleventh Symposium (International) on Combustion*, pp.791-798, The Combustion Institute, 1967.
53. Batchelor, G.K. "Smale Scale Variation of Convected Quantities Like Temperature in Turbulent Fluid", *J. Fluid Mech.* **5**, 113-133, 1959.
54. Grandmaison, E.W. "Turbulent Mixing in Free Swirling Jets", Ph.D. Thesis, Queen's University, Kingston, Ontario, 1975.
55. Becker, H.A., Hottel, H.C. and Williams, G.C. "Concentration Intermittency in Jets", *Tenth Symposium (International) on Combustion*, pp.1253-1263, The Combustion Institute, 1965.
56. Wygnanski, I. and Fiedler, H. "Some Measurements in the Self-preserving Jet", *J. Fluid Mech.* **38**, 577-612, 1969.
57. Williams, G.C., Hottel, H.C. and Gurnitz, R.N. "A Study of Premixed Turbulent Flames by Scattered Light", *Twelfth Symposium (International) on Combustion*, pp.1081-1092, The Combustion Instititе, 1969.
58. Danckwerts, P.V. "The Definition and Measurement of Some Characteristics of Mixtures", *Appl. Sci. Res.* **A3**, 279-296, 1953.
59. Hawthorne, W.R., Weddell, D.S. and Hottel, H.C. "Mixing and Combustion in Turbulent Gas Jets", *Third Symposium on Combustion, Flame and Explosion Phenomena*, pp. 266-288, Williams and Wilkins, 1949.
60. Becker, H.A. and Booth, B.D. "Mixing in the Interaction Zone of Two Free Jets", *A.I.Ch.E.J.* **21**, 949-958, 1975.
61. Favre, A. "Statistical Equations of Turbulent Gases". *In* "Problems of Hydrodynamics and Continuum Mechanics", pp.231-266, Society for Industrial and Applied Mathematics, Philadelphia, Pennsylvania, 1969.
62. Bilger, R.W. "A Note on Favre Averaging in Variable Density Flows", *Comb. Sci. & Tech. (in press).*
63. Bilger, R.W. "Turbulent Jet Diffusion Flames", *In* "Progress in Energy and Combustion Science, **Vol.I**", Pergamon, 1975.
64. Bilger, R.W. "The Structure of Diffusion Flames", *Comb. Sci. & Tech. (in press).*
65. Monin, A.S. and Yaglom, A.M. "Statistical Fluid Mechanics", MIT Press, 1971.
66. Tennekes, H. and Lumley, J.L. "A First Course in Turbulence", MIT Press, 1972.
67. Spalding, D.B. "Mathematical Models for Free Turbulent Flows", *Istituto Nazionale di Alta Matematica Symposia Matematica*, **Vol.IX**, pp.391-416, 1972.
68. Harlow, F.H. "Turbulence Transport Modelling", AIAA Selected Reprint Series, **Vol.XIV**, AIAA, 1973.
69. Hinze, J.O. "Turbulence", 2nd ed., McGraw-Hill, 1975.
70. Williams, F.A. "Combustion Theory", Addison Wesley, 1965.
71. Spalding, D.B. "Concentration Fluctuations in a Round Turbulent Free Jet", *Chem. Eng. Sci.* **26**, 95-107, 1971.
72. Becker, H.A., Hottel, H.C. and Williams, G.C. "Mixing and Flow in Ducted Turbulent Jets", *Ninth Symposium (International) on Combustion*, pp.7-19, The Combustion Institute, 1963.

73. Hottel, H.C. and Sarofim, A.F. "The Effect of Gas Flow Patterns on Radiative Transfer in Cylindrical Furnaces", *Int. J. Heat Mass Transfer* **8**, 1153-1169, 1965.
74. Becker, H.A. and Brown, A.P.G. "Velocity Fluctuations in Turbulent Jets and Flames", *Twelfth Symposium (International) on Combustion*, pp.1059-1068, The Combustion Institute, 1969.
75. Bilger, R.W. and Kent, J.H. "Concentration Fluctuations in Turbulent Jet Diffusion Flames", *Comb. Sci. Tech.* **9**, 25-29, 1974.
76. Rathgeber, D.E. "Mixing Between a Round Jet and a Transverse Turbulent Pipe Flow", Ph.D. Thesis, Queen's University, Kingston, Ontario, 1974.
77. Clemons, D.B.
Ph.D. Dissertation, Princeton University, 1962.
78. Kim, Y.G.
Ph.D. Dissertation, Princeton University, 1963.
79. Christiansen, D.E. "Turbulent Liquid Mixing", *I. & E.C. Fund.* **8**, 263-271, 1969.
80. Goulard, R. "Laser Raman Scattering Applications", *J.Q.S.R.T.* **14**, 969-974, 1974.
81. Hartley, D.L. "Transient Gas Concentration Measurements Utilizing Laser Raman Spectroscopy", *AIAA J.* **10**, 687-689, 1972.
82. Hartley, D.L. "Application of Laser Raman Spectroscopy to the Study of Factors that Influence Turbulent Gas Mixing Rates", *In* "Fluid Mechanics of Mixing", pp.131-146, American Society of Mechanical Engineers, 1973.
83. Setchell, R.E. "Analysis of Flame Emissions by Laser-Raman Spectroscopy", *Western States Section, Spring Meeting*, Paper No. WSS/CI 74-6, The Combustion Institute, 1974.
84. Corrsin, S. and Uberoi, M. "Further Experiments on the Flow and Heat Transfer in a Heated Turbulent Jet", NACA Rept. 998, 1950; NACA TN 1865, 1949.
85. Wilson, R.A.M. and Danckwerts, P.V. "Studies in Turbulent Mixing – II, a Hot-air Jet", *Chem. Eng. Sci.* **19**, 885-895, 1964.
86. Antonia, R.A. and Bilger, R.W. "The Heated Round Turbulent Jet in a Coflowing Stream", Charles Kolling Research Laboratory TN F-66, Dept. of Mechanical Engineering, University of Sydney, Sydney, Australia, 1974.
87. Way, J. and Libby, P.A. "Applications of Hot-wire Anemometry and Digital Techniques to Measurements in a Turbulent Helium Jet", AIAA Paper No. 71-201, 1971.
88. McQuaid, J. "Turbulence Measurements with Hot-wire Anemometry in a Nonhomogeneous Jet". *In* "Fluid Mechanics of Mixing", American Society of Mechanical Engineers, 1973.
89. Antonia, R.A., Prabhu, A. and Stephenson, S.E. "Conditionally Sampled Measurements in a Heated Turbulent Jet", Charles Kolling Research Laboratory TIV F-72, Dept. of Mechanical Engineering, University of Sydney, Sydney, Australia, 1974.
90. Antonia, R.A. and Prabhu, A. "Reynolds Shear Stress and Heat Flux Balance in a Turbulent Round Jet", *AIAA J. (in press).*
91. Richardson, J.M., Howard, H.C. and Smith, R.W. "The Relation Between Sampling-tube Measurements and Concentration Fluctuations in a Turbulent Gas Jet", *Fourth Symposium (International) on Combustion*, p.814, Williams and Wilkins, 1953.
92. Toor, H.L. "Mass Transfer in Dilute Turbulent and Non-turbulent System with Rapid Irreversible Reaction and Equal Diffusivities", *A.I.Ch.E.J.* **8**, 70-78, 1962.
93. Libby, P.A. "On Turbulent Flows with Fast Chemical Reactions. Part III: Two-dimensional Mixing with Highly Dilute Reactants", *Comb. Sci. & Tech. (in press).*
94. Corrsin, S. "The Reactant Concentration Spectrum in Turbulent Mixing with a First-order Reaction", *J. Fluid Mech.* **11**, 407-416, 1961.
95. Vassilatos, G. and Toor, H.L. "Second Order Chemical Reactions in a Nonhomogeneous Turbulent Fluid", *A.I.Ch.E.J.* **11**, 666-672. 1965.
96. Toor, H.L. "Turbulent Mixing of Two Species with and without Chemical Reactions", *Ind. Eng. Chem. Fund.* **8**, 655-659, 1969.
97. Mao, K.W. and Toor, H.L. "Second-order Chemical Reaction with Turbulent Mixing", *Ind. Eng. Chem. Fund.* **10**, 192-197, 1971.
98. Yieh, H.
Ph.D. Thesis, Ohio State University, 1971.
99. Patterson, G.K. "Model with No Arbitrary Parameters for Mixing Effects on Second-order Reaction with Unmixed Feed Reactants", *In* "Fluid Mechanics of Mixing", The American Society of Mechanical Engineers (1973).

100. Borghi, R. "Chemical Reaction Calculations in Turbulent Flows", *Second IUTAM-IUGG Symposium on Turbulent Diffusion in Environmental Pollution*, Charlottesville, Virginia, 1973.
101. Borghi, R. "Computational Studies of Turbulent Flows with Chemical Reaction", ONERA, Chatillon, 1974.
102. Stewart, W.E. and Prober, R. "Matrix Calculation of Multicomponent Mass Transfer in Isothermal Systems", *Ind. Eng. Chem. Fund.* 3, 224-235, 1964.
103. Von Behren, G.L., Jones, W.O. and Wasan, D.T. "Multicomponent Mass Transfer in Turbulent Flow", *A.I.Ch.E.J.* **18**, 25-30, 1972.
104. Stewart, W.E. "Multicomponent Mass Transfer in Turbulent Flows", *A.I.Ch.E.J.* **19**, 398-400, 1973.
105. Lapedes, D.N. (Ed.-in-Chief) "McGraw-Hill Dictionary of Scientific and Technical Terms", McGraw-Hill, 1974.
106. Weiser, H.B. "A Textbook of Colloid Chemistry" (2nd ed.), Wiley, 1958.
107. Bradshaw, P. "Experimental Fluid Mechanics", Macmillan, 1964.
108. Pankhurst, R.C. and Holder, D.W. "Wind-Tunnel Technique", Pitman, 1952.
109. Maltby, R.L. and Keating, R.F.A. "Flow Visualization in Low-speed Wind Tunnels", Tech. Note AERO. 2715, Royal Aircraft Establishment (Bedford), Ministry of Aviation, London, 1960.
110. Nolen, J.T. "Particle Size of Smokes from Induction Nozzles", Ph.D. Thesis, Massachusetts Institute of Technology, Cambridge, Massachusetts, 1946.
111. Green, H.L. "Particulate Clouds: Dusts, Smokes, and Mists", Van Nostrand, 1964.
112. Kuhn, W.E. (Ed.-in-Chief) "Ultrafine Particles", Wiley, 1963.
113. Koldschütter, V. and Tüscher, J.L. "Über darstellung disperser Substanzen in Gasförmigen Medien", *Z. f. Elektrochemie* **27**, 225-256, 1921.

NOMENCLATURE

Signs

$(\ )^*$	quantity associated with marker particles
$\tilde{(\ )}$	molar value, Section 5
$<(\ )>$	average value
$\overline{(\ )}$	time-average value
$\hat{(\ )}$	$\equiv \sqrt{\overline{(\ )^2}}$, root-mean-square value
$\breve{(\ )}$	volume-average value within the probe control volume

Subscripts

a	value for absorption of radiation
b	value for background radiation
C	value for any chemical component or complex of components introduced with a given feed
esn	value for electronic shot noise
f	value at frequency f
i,j	molecular species
k	element (atomic species)
L	value associated with light source
l,m,n	physicochemically distinct feed materials
msn	value for marker shot noise
o	value in a feedstream
p	particle property
P	value associated with phototransducer
s	value for scattering of radiation

Variables

Note: Variables appearing in a given meaning in only one section are defined in context.

A_p projected area of particle, m^2

$A_{\lambda a}$ spectral absorption cross-section of particle, m^2

$A_{\lambda s}$ spectral scattering cross-section of particle, m^2

a_i radius of incident laser beam at the $1/e^2$ point, m

$\mathcal{C}$ number concentration of marker particles, m^{-3}

c molar gas density, mol/m^3

C_p heat capacity, J/kg K

D_i effective diameter of incident beam over the pcv, m

D_p marker particle diameter, m

D_V effective diameter of pcv, calculated as diameter of a sphere with the same volume, m;

$D_V = (6V/\pi)^{1/3}, \quad V = \pi D_i^2 L_i/4$

$\mathcal{D}$ diffusivity of molecular species in fluid, m^2/s

$\mathcal{D}^*$ diffusivity of marker particles in fluid, m^2/s

$E(\kappa)$ wave-number one-dimensional spectral density function of molecular species concentration fluctuations γ, kg^2/m^5

$E^*(\kappa)$ wave-number spectral density function of marker concentration fluctuations γ^*, kg^2/m^5

$E_V^*(\kappa)$ wave-number spectral density function of volume-averaged marker concentration fluctuations γ^*, kg^2/m^5

e_K $\equiv \frac{1}{2}\overline{\mathbf{u}\cdot\mathbf{u}}$, specific turbulence kinetic energy, m^2/s^2

f frequency, s^{-1}

$F(\tau)$ residence time distribution function of marker particles in the pcv

G_i flux density of radiation from the incident beam on a plane oriented normal to the beam and containing the centre-point of the pcv, W/m^2

$G_{\lambda i}$ spectral value of G_i, W/m^3

$G(f)$ frequency spectral density function of molecular species concentration fluctuations γ, $kg^2\,s/m^6$

$G^*(f)$ frequency spectral density function of marker concentration fluctuations γ^*, $kg^2\,s/m^6$

$G_V^*(f)$ frequency spectral density function of volume-averaged marker concentration fluctuations γ^*, $kg^2\,s/m^6$

g_{lm} $\equiv \overline{\gamma_l \gamma_m}/\overline{\Gamma}_l\overline{\Gamma}_m = \overline{y_l y_m}/\overline{Y}_l\overline{Y}_m$, index of molecular segregation between material from feeds l and m

$H(f)$ frequency spectral density function of phototransducer output current fluctuations i, $A^2\,s$

$H_{esn}(f)$ frequency spectral density function of output current fluctuations due to electronic shot noise, A^2 s

$H_L(f)$ frequency spectral density function of output current fluctuations due to incident beam intensity fluctuations, $A^2\,s$

$H_{msn}(f)$ frequency spectral density function of output current fluctuations due to marker shot noise, $A^2\,s$

h	heat transfer coefficient between fluid and particles, $W/m^2 K$
h	specific enthalpy of fluid, J/kg
h_i	specific enthalpy of molecular species i, J/kg
I_λ	spectral intensity of radiation, W/m^2 sr
I	phototransducer output current, A
I^*	output current response to marker concentration at the measurement point in the absence of volume averaging and extinction along the optical path, A
I_b	output current due to radiative background, A
I_d	phototransducer dark current, A
$I_s{}^*$	output current response to the concentration field in the pcv, A
i	$\equiv I - \bar{I}$, output current fluctuation, A
i^*	fluctuation in I^*, A
i_b	fluctuation in I_b, A
i_{esn}	fluctuation in I due to electronic shot noise, A
i_E	fluctuation in I due to fluctuations in incident beam intensity, A
i_s^*	fluctuation in I_s^*, A
$\mathfrak{J}$	source function
$\mathbf{j}_i$	diffusion mass flux vector for molecular species i, kg/m^2 s
K_λ	total spectral extinction coefficient, m^{-1}
$K_{\lambda a}$	spectral absorption coefficient, m^{-1}
$K_{\lambda s}$	spectral scattering coefficient, m^{-1}
K_g	Cunningham factor
k	thermal conductivity of fluid, W/m K
k_p	mass transfer coefficient based on partial pressure driving force, $(kg/m^2 s)(m^2/N)$
k_r	coagulation rate constant, m^3/kg s
L	length of optical path, m
L_i	optical depth of pcv with respect to incident beam, m
L_m	maximum pathlength of particles traversing the pcv, m
ℓ	molecular mean free path, m
Le	Lewis number
m	$\equiv n(1 + j\kappa)$, complex index of refraction
m_p	particle mass, kg
M	molecular weight, kg/mol
N	number of particles in pcv
N^*	$\equiv \Gamma^* V/m_p$, number of particles in pcv corresponding to concentration at centrepoint
n	real index of refraction
P	total pressure, N/m^2
p_λ	phase function for scatter of radiation
$p_{\lambda s}^*$	phase function for scatter at angle θ_s to the incident beam
pcv	probe control volume
Q	volume-averaging factor on the phototransducer current response to marker concentration fluctuations in the pcv in a given frequency interval

Q_f value of Q at a given frequency f, applicable to the frequency spectral density function of the fluctuations

Q_κ value of Q at a given wave number $\kappa = 2\pi f/U$, applicable to the wave-number spectral density function

R distance along a beam of radiation, m

$R(\mathbf{x}_A;\mathbf{x}_B) \equiv \overline{\gamma_A \gamma_B}/\hat{\gamma}_A \hat{\gamma}_B$, two-point correlation function for concentration fluctuations

$R_V(\mathbf{x}_A;\mathbf{x}_B) \equiv \overline{\breve{\gamma}_A \breve{\gamma}_B}/(\overline{\breve{\gamma}_A{}^2}\,\overline{\breve{\gamma}_B{}^2})^{1/2}$, two-point volume-averaged correlation function for concentration fluctuations

R_{lm} $\equiv \overline{\gamma_l \gamma_m}/\hat{\gamma}_l \hat{\gamma}_m$, correlation function for concentration fluctuations of material from two feeds

r radial coordinate in a cylindrical polar coordinate system; radial position in the nephelometer incident beam within the pcv, m

r_i mass rate of chemical reaction per unit volume, $kg/m^3 s$

S control surface, m^2

S_i a control plane passing through the centre of the pcv normal to the axis of the incident beam, m^2

S_p surface area of particle, m^2

$s_{\lambda P}$ current sensitivity of phototransducer to radiation at wavelength λ, A/W

s^* current sensitivity of phototransducer to marker concentration, $A\ m^3/kg$

s_p^* current sensitivity of phototransducer to marker concentration for particles of diameter D_p, $A\ m^3/kg$

Sc $\equiv \nu/D$, Schmidt number for molecular species

Sc* $\equiv \nu/D^*$, Schmidt number for marker particles

T_λ spectral transmittance; the radiant flux transmitted by a system divided by the incident radiant flux at the given wavelength λ

$T_{\lambda D}$ spectral transmittance of the detector optics

$T_{\lambda E}$ spectral transmittance of the exciter optics

$T_{\lambda L}$ spectral transmittance of the lamp window

$T_{\lambda N}$ $\equiv T_{\lambda E} T_{\lambda D}$, total spectral transmittance of nephelometer optical components

$T_{\lambda W}$ spectral transmittance of windows in apparatus walls, etc.

$T_{\lambda i}$ spectral transmittance of the path between the exciter and the pcv

$T_{\lambda s}$ spectral transmittance of the path between the detector and the pcv

T temperature, K

T_p particle temperature, K

$\mathbf{U}$ stream velocity vector, m/s

U magnitude of mean stream velocity vector, m/s

U_x, U_y, U_z velocity components in a Cartesian coordinate system

$\mathbf{u}$ $\equiv \mathbf{U} - \overline{\mathbf{U}}$, fluctuating component of $\mathbf{U}$

u_x, u_y, u_z fluctuating components of U_x, U_y and U_z, m/s

$\mathbf{U}^\dagger$ molar-average transport velocity, m/s

$\mathbf{U}^\ddagger$ volume-average transport velocity, m/s

$\mathbf{U}_i$ molecular species transport velocity, m/s

$\mathbf{v}$	particle velocity, m/s
$\mathbf{v}_t$	terminal particle velocity, m/s
V	molar volume, m^3/mol
V_i	partial molar volume of species i, m^3/mol
W	mass fraction of a given feed
W_C	mass fraction of a given component or complex of components from a given feed
W_l, W_m, W_n	mass fractions of material from feeds l, m and n
W_i, W_j	mass fractions of molecular species i and j
W_k	mass fraction of atomic species k
w_l, w_m, w_n	fluctuations in W_l, W_m and W_n
w_i, w_j	fluctuations in W_i and W_j
w_k	fluctuation in w_k
$\mathbf{x}_A, \mathbf{x}_B$	position vectors of points A and B, m
x, y, z	Cartesian coordinates, m
X_i, X_j	mole fractions of species i and j
Y_l, Y_m, Y_n	$\equiv \Gamma_l/\Gamma_{l,o}, \Gamma_m/\Gamma_{m,o}, \Gamma_n/\Gamma_{n,o}$; mass dilution ratios
y_l, y_m, y_n	fluctuations in Y_l, Y_m, Y_n
Z_i	$\equiv r_i/\rho$, specific mass rate of chemical reaction, s^{-1}
α	$k/\rho C_p$, thermal diffusivity, m^2/s
β	fractional fluctuation in light source radiant intensity
Γ	mass concentration of a given molecular species, kg/m^3
Γ^*	mass concentration of marker particles, kg/m^3
γ	fluctuation in Γ, kg/m^3
γ^*	fluctuation in Γ^*, kg/m^3
$\overline{\delta}$	concentration intermittency function, $\delta = 0$ when $\Gamma = 0$, $\delta = 1$ when $\Gamma > 0$
δ	$\equiv \text{prob}(\Gamma > 0) \equiv \text{prob}(\delta = 1)$, intermittency factor
ϵ_T	rate of dissipation of turbulence kinetic energy per unit mass, m^2/s^3
ϵ_i	volume fraction of molecular species i
ζ	$\equiv \lvert \mathbf{x}_A - \mathbf{x}_B \rvert$ separation distance of two points, m
ζ_i, ζ_j	fluctuations in Z_i, Z_j
η	total extinction efficiency for extinction of radiation by a particle
η_a	absorption efficiency
η_s	scatter efficiency
η_λ	spectral extinction efficiency
$\eta_{\lambda a}$	spectral absorption efficiency
$\eta_{\lambda s}$	spectral scatter efficiency
η_E	collection efficiency of exciter optics; fraction of radiation intercepted from light source
θ	scatter angle, angle between a given direction of scatter and the incident beam, rad
θ_s	angle of scatter used in a given measurement with a nephelometer, rad
θ	temperature fluctuation, K
κ	$\equiv 2\pi f/\overline{U}$, wave number, m^{-1}

Λ	integral length scale of turbulence, m
Λ_λ	integral length scale for concentration fluctuations, m
Λ_u	integral length scale for velocity fluctuations, m
λ	wave length of radiation, m
λ_u	Kolmogoroff microscale of velocity fluctuations, m
μ	viscosity of fluid, kg/m s
ν	kinematic viscosity of fluid, m^2/s
ν_T	kinematic eddy viscosity, m^2/s
ρ	density of fluid, kg/m^3
ρ_p	particle density, kg/m^3
σ	Stefan-Boltzman constant, $W/m^2 K^4$
τ	optical transmissivity along a path through a medium, fraction of incident radiation not extinguished from a beam in the given direction by absorption or scatter
τ_λ	spectral transmisivity
$\tau_{\lambda L}$	spectral transmissivity along a path of length L
Φ	radiant energy flux, *W*
Φ_λ	spectral radiant energy flux, W/m
$\Phi_\lambda{}^*$	spectral flux received by the detector if the marker concentration were uniformly Γ^* throughout the pcv, if there were no extinction between the pcv and the detector, and if there were no marker shot noise, W/m
$\Phi^*_{\lambda s}$	spectral marker-scattered radiant flux received by the detector, W/m
$\Phi_{\lambda E}$	spectral radiant flux in the incident beam leaving the exciter, W
$\Phi_{\lambda P}$	spectral radiant flux received by the phototransducer, W
Ω	solid angle, sr
Ω_i	solid angle included by incident beam, sr
Ω_s	solid angle through which scattered light is received from the pcv by the detector, sr
Ω_{lm}	$\equiv \overline{w_l w_m}/\hat{w}_l \hat{w}_m$ correlation function for fluctuations in the mass fractions of material from feeds l and m at a given point
ω_n	$\equiv \overline{w_n{}^2}/\overline{W}_n(1 - \overline{W}_n)$, an intensity of segregation

APPENDIX

NOTES ON APPARATUS FOR MARKER NEPHELOMETRY

A.1 General

A detection and measurement system for marker nephelometry consists of the following subsystems:

1. An exciter – the light source and associated optics for producing the incident light beam.
2. A detector – the receiving optics and a photoelectric transducer. The exciter and detector together constitute the nephelometer probe.
3. A traversing system for moving the pcv around in the field under investigation.
4. A sol generator; for gases, a "smoke" generator.
5. Instrumentation for signal processing, recording and display.

It is not intended here to give a comprehensive description of any complete system, but simply to indicate some suitable components and to point out some critical design features and operating conditions.

A.2 The Exciter

The operating characteristics normally required of a light source are:

1. Stability. Any drift in output power should be very gradual. Drift of less than 1% in 8 h is attainable in light-regulated lasers.
2. Low fluctuation noise. If it is desired to measure concentration fluctuations, then fluctuations in the incident beam power should be relatively small. Root mean square fluctuation levels below 3% of the mean beam power in the frequency band 0–50 kHz can be achieved with any of the light sources mentioned below, and levels under 0.5% are attainable with some. The minimum fluctuation noise for a given lamp or laser is ultimately determined by its power supply, and care should be taken (when there is a choice) to select one with low output ripple. The effect of source fluctuation noise can usually be corrected for by calibration, as indicated by (3.8.1) and (3.8.2).

The following light sources have been used for air/air mixing studies: 100 W concentrated arc lamps [19–21,43,44,51,52,55,72], 75 W xenon arc lamps [60], 5 mW He-Ne lasers [45,46,51), and a 2 W argon ion laser [23,48]. A 0.6 W light-regulated argon ion laser is in use in the author's laboratory in work on flames, the high power being required to bring the scattered light intensity well above the background due to flame radiation. Liu and Karaki [47] used a 750 W tungsten projector lamp in their very compact nephelometer probe.

Typical exciter optics for an arc lamp are shown in Fig.4. The light-gathering efficiency η_L is largely determined by the ratio D_1^2/x_1^2, where D_1 is the diameter of the front aperture stop A.S. The lens L_1 produces an image of the arc on a diaphragm whose aperture, of diameter D_2, acts as a field stop. The lens L_2 produces the probe incident beam. The pcv is normally located at the point of

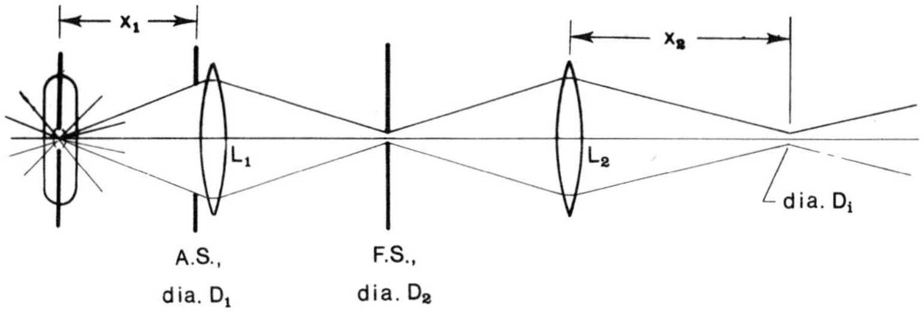

Fig.4 Exciter optics for a system with an arc lamp as the light source.

focus, and its diameter D_i can be changed by using a different field stop diameter D_2. The lenses L_1 and L_2 may in fact be combinations of lenses. For example, the author has used a pair of plano-convex lenses for L_1 and a coated anastigmat projection lens system for L_2 [60]. A lens system L_1 is usually provided with commercially available lamp housings. The distance x_2 is determined by the clearance required between L_2 and the field under study. The author has used lenses of 3–8 cm dia, x_2 from 10 to 50 cm, and pcv diameters D_i from 0.5 to 2.3 mm (though most commonly $D_i \approx 1$ mm).

Beam modifying optics used with lasers are usually very simple, and are available as accessories from several laser manufacturers. The beam may be expanded, collimated, or focussed, or combinations of these operations may be done.

Lasers are usually preferable to lamps, because their energy is concentrated at a few principle wavelengths and in a narrow beam of very small angle of divergence. However, lamps do provide a broader selection of usable wavelengths, should that be desirable. The concentrated arc in particular has very nearly grey-body emission characteristics.

A.3 The Detector

Typical detector optics are shown in Fig.5. The input of scattered light from the pcv is determined by D_1^2/x_1^2 – or, basically, by the solid angle $\Omega_s \approx \pi D_1^2/4x_1^2$, where D_1 is the diameter of the front aperture stop or entrance pupil. Light scattered along the path of the incident beam is focussed by the lens L on the

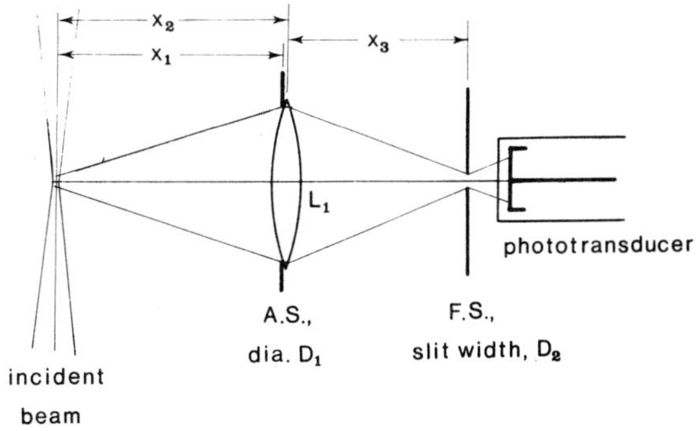

Fig.5 Detector optics.

slitted diaphragm F.S. which acts as the field stop, and the slit width D_2 determines the pcv length (or optical depth) with respect to the incident beam, L_i. The minimum distance x_1 is determined by the clearance requirements of the experiment. The author has usually operated with lenses L_1 of 5–8 cm dia, x_1 from 30 to 30 cm, and $x_2 \approx x_3$. Control volume dimensions were usually such that $L_i \approx 0.8\ D_i$.

The lens L_1 may in fact be a combination of lenses. The author has usually used pairs of convex or planoconvex simple lenses. The detector optics may include other elements, most commonly optical filters placed between the slit and the phototransducer. Narrow-band interference filters have been used in all work with lasers. In work with flames a sheet of heat absorbing glass has been placed before the lens, and the detector housing has been insulated on the outside and air-cooled on the inside. In the system currently in use the scattered-light signal is received by a fibre-optic line which transmits it to a photomultiplier tube 2 m away and outside the test chamber. The fibre-optic bundle has a roughly 1 mm by 10 mm rectangular or "flat" termination at the receiving end which is utilized as the field stop or slit.

The most suitable photoelectric transducers are photomultiplier tubes and silicon photodiodes, and a good selection of these is available. The characteristics are usually well described in manufacturer's literature. Photomultiplier tubes have the advantage that they usually do not require a preamplifier to boost the output signal. They can also provide very large sensor (photocathode) areas, should that be a consideration. Photodiodes are simple, rugged, and very small – typically, 0.5–1 cm dia. However, their sensor areas may be too small for some applications, and they usually need a preamplifier to raise their output to the level required by most measuring instruments.

The spectral sensitivity of a photoelectric transducer, $s_{\lambda P}$, was introduced in Sec.3.4; it is the output current (amperes) per unit of input radiant flux (watts). The spectral sensitivity of the most common types of photomultiplier tubes (SbCs or SbKCs cathode, glass window) is illustrated by the following example of a tube used by the author:

1. Maximum at 425 nm.
2. About 88% of maximum at 488.0 nm, and 65% at 528.7 nm, these being the principal argon ion laser lines.
3. About 3% of maximum at 632.8 nm, the principle He-Ne laser line.

The peak cathode quantum efficiency (electrons per photon) of such tubes is typically 0.15–0.28.

Photomultiplier tubes designed for infrared work typically have maximum sensitivity around 800 nm, and about 80% of maximum at 632.8 nm. Their maximum quantum efficiency is similar to that of the ordinary types described above. Their cathode dark current may, however, be around 500 times higher.

A number of silicon photodiodes are available whose sensitivity peaks around 800 nm, and is over 50% of the maximum from 450 nm to 1000 nm. Their quantum efficiency is 0.4–0.8 over this range. The cathode dark current of some types is remarkably low. These devices would be significantly superior to standard photomultipliers for use with He-Ne lasers in cases where the photomultiplier has unfavourable signal/noise ratios at the available levels of irradiation, e.g., the electronic shot noise level $\overline{i^2_{esn}}/\bar{I}^2$ would be reduced by a factor as large as 30.

The author's work, however, has nearly all been done with standard 9-stage or 10-stage photomultiplier tubes. When he changed light-sources in work with flames from a 5 mW He-Ne laser to a 250 mW line of an argon-ion laser, the photomultiplier output changed by a factor of 50 from the change in power, and a further factor of 30 from the shift in radiation wavelength to a more sensitive region of the photomultiplier response, giving a total factor of 1500; this estimate, of course, neglects the probable gain in light scatter.

Photoelectric transducers should be worked, when possible, near the upper end of their linear range, insofar as this is necessary to keep electronic shot noise and dark current relatively low. To prevent current-induced nonlinearity, the current in the voltage divider circuit of a photomultiplier tube should generally be at least 100 times the output (anode) current. This means that energy must be dissipated in the voltage divider at rates as high as 10 W or more. Care should therefore be taken that the resistors in the voltage divider have an adequate power rating and that the divider is adequately ventilated. Photomultiplier tube housings complete with voltage dividers are easy to build. They are also available commercially, but care should be taken to ascertain that the specifications of such units are adequate.

The power supply for the phototransducer should, for careful work, have very low drift and ripple.

The signal resistor, through which the output current of the phototransducer produces the observed voltage signal, should have an adequately low value for good signal transmission to the signal processing system. The resistor and the transmission line constitute an RC filter, and to keep the frequency response up the time constant RC must be small, preferably under 10 μs. The line capacitance C is kept small by using a good coaxial shielded cable and by keeping the line short. It should be remembered that photoelectric devices are generally current generators, not voltage generators, and this results in different signal transfer characteristics.

A.4 The Nephelometer Probe and the Traversing Arrangements

The exciter and the detector, which together constitute the nephelometer probe, are generally coupled mechanically or are mechanically maintained in some fixed relation by the traversing system. If they are rigidly coupled, then the detector always sees the same segment of the incident beam, and the gathered radiation always falls on the same area of the cathode in the phototransducer. An attractive alternative in some situations, especially with a laser light source, is to conduct traverses in one dimension by moving the detector parallel to the incident beam. Equation (3.3.10) shows that, except for the effect of volume-averaging, the detector response is determined solely by the pcv's incident optical depth L_i, and is insensitive to changes in the beam diameter.

The exciter/detector coupling and the field traversing arrangements depend very much on the experiment, and each new problem tends to have a different optimum solution. In air/air mixing studies by the author and coworkers involving pipe flows [51,76], the pipe diameters were 20 cm or less. The incident beam was there held stationary with respect to the pipe, directed through the centre-line at right-angles. Diametrical traversing was done by scanning the incident beam, by moving the detector parallel to it. Longitudinal traversing was done by moving the smoke source (a jet nozzle [76] or iso-kinetically operated, axially directed, tubular injector [51]) to different stations along the pipe wall. In the case of the transverse jet [76] the field was three-dimensional, and a polar traverse was done by rotating the section of pipe to which the nozzle was fitted. This approach – with a stationary light source and a traversed detector – is best carried out with a photomultiplier tube providing a large cathode area, so that small misalignments will not significantly alter the response as the detector is moved along. The work mentioned was done with a concentrated arc lamp in one case [51], and a He-Ne laser in the other [76].

In other work by the writer and his colleagues, exciter and detector have been rigidly coupled together, and always operated in such a fashion as to be effectively clear of the field under study, so that the probe's presence in the field was purely optical.

The traversing rigs used in the author's laboratory have all been made from commercially available components: triangular rails and attachments from optical apparatus suppliers, and traverse tables (with single, compound, or rotary motion), post drill stands, and lathe-type traverse beds from machine tool suppliers.

A novel arrangement was developed to fit the demands of the work with free turbulent diffusion flames, Fig.6. The flame was inside a ventilated cell. The exciter (an argon ion laser) and the photomultiplier tube were outside, together with the operator and all instruments. The light signal was brought to the photomultiplier tube by a fibre optic transmission line, as mentioned earlier. The detector optics were mounted in a housing on the end of an arm A made of aluminium tubing. The support arm and the laser were both mounted on the rotatable table of a machine drill stand D.S. Horizontal traversing was done by rotating the table, swinging the pcv through an arc intersecting the flame jet axis. Horizontal displacement was measured by means of a dial gauge D.G. with 15 cm maximum travel acting against a fixed vertical post F.V.P. made of precision-ground steel rod.

There may be experiments where the field under study is so large, or is so inaccessible, that the nephelometer probe must enter it mechanically, not just optically. It may also be desirable that the nephelometer be portable for use in "field" experiments or tests. The compact probes of Yang and Meroney [45,46] and Liu and Karaki [47] were designed to meet such requirements.

One may also wish to make measurements under conditions where there are severe problems with signal modulation by extinction along the light path, and/or optical background noise. Langer and coworkers [43,44] solved this problem in work on dust-laden jets by bringing the incident beam in, and taking the scattered beam away, through glass tubes (light pipes) extending into the field up to the pcv. Probes designed on the same principle as Liu and Karaki's [48] should often be ideal for such situations. In their arrangement the exciter and detector front

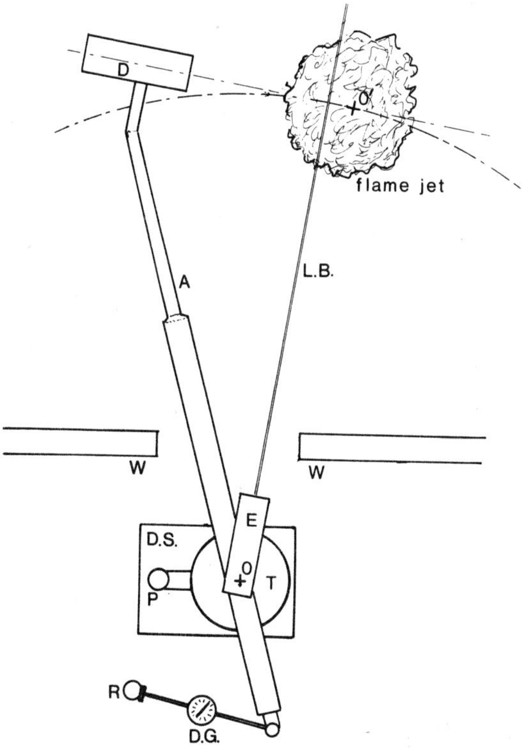

Fig. 6 **Traversing system for studies of free jet flames. E, exciter (an argon-ion laser); L.B., laser beam; D, detector optical box; W, walls of test cell; D.S., machine drill stand; P, drill-stand vertical traverse post; T, drill-stand table; R, precision-ground rod; D.G., dial gauge. The laser E and the detector support arm A are mounted on the rotatable table T. Scattered light gathered by the optical system in the box D is fibre-optically transmitted down the arm A to a photomultiplier tube outside the cell.**

ends are both brought close to the pcv, facing each other, the exciter end blocks the detector's view of the background, but the incident light-beam is blocked in the detector. The scattered light is taken from the forward direction at very small scattering angles θ_s, effectively in the region of maximum scattering intensity.

A.5 Smoke Generators

Smokes have been defined [105] as aerosols whose particles are in the diameter range 0.01–5 μm. The particles may be solid, or they may be liquid droplets. In the latter case it is generally understood that the liquid's vapour pressure is very low, so that the droplets have a fair degree of permanence.

Fogs are dispersions whose particles are liquid droplets formed by condensation of vapour in a vapour-gas mixture. By this definition, smokes of liquid droplets are also fogs. However, the term is usually applied to droplets of liquids with an appreciable vapour pressure. Natural water fog droplets [106] are in the diameter range 4–30 μm. The term mist is meteorologically applied to thin fogs in which horizontal visibility is over 1 km.

The colloidal size range has been defined [105] has 1 nm–1 μm. On this basis some smokes would be classified as colloidal dispersions and others as coarse dispersions (dia > 1 μm). However, smoke particles generally exhibit a lively Brownian motion and a strong Tyndall effect, and both fogs and smokes are customarily spoken of as aerosols.

A variety of fog and smoke generators have been developed for flow visualization studies. A number of authors [107–109] give designs for the Preston-Sweeting kerosene mist generator (better called a fog generator) and variants thereof. In this device kerosene is boiled in a tube, and the vapour issues as a jet through a tiny nozzle or orifice into a stream of cold air with which it jet-mixes and in which it condenses, forming a white fog. This fog would not usually be suitable for marker nephelometry, the vapour pressure of kerosene being considerably too high to provide stable particles unless all the gas in the system is saturated with kerosene. The particles may also be too large and their number concentration too low.

Whytlaw-Gray and Patterson [39] made stearic acid smokes by evaporating stearic acid into a cold air stream. Nolen [110] encountered difficulties in devising a more sophisticated stearic acid smoke generator, and instead developed a generator for oil condensation smokes. He used an oil called Diol 55, made by the Standard Oil Co. The oil was pumped through a preheater into a vaporizing chamber made of steel pipe 10 cm dia by 50 cm high. The vaporizer was wrapped with insulation and chromel electrical heating wire. The electrical power input was such as to supply the heat losses and regulate the rate of vaporization. The vapour line leaving the vaporizer was electrically heated to prevent condensation. The vapour was mixed with a hot air stream. The air-vapour mixture was then discharged through a flow nozzle or orifice into a stream of cold air. The oil condensed in the jet mixing region, forming a dense smoke. Air/vapour temperatures at discharge were 237°C and 342°C, discharge velocities were 30–200 m/s, and diol mass fractions were 0.05–0.5. Flow nozzles of 0.32–0.64 cm dia were used, and a sharp-edged orifice of 0.22 cm dia.

Nolen determined mean, minimum, and maximum smoke particle diameters. The author has found that the following empirical formula gives a fairly good representation of the results for the flow nozzles:

$$D_p = 1140\,(W_o U_o D_o)^{0.6} / T_o \ , \qquad \text{(A.5.1)}$$

where the mean particle diameter D_p is in μm, diol mass fraction W_o is dimensionless, discharge velocity U_o is in m/s, nozzle diameter D_o in m, and discharge temperature in K. The results for the orifice plate are ambiguous; the best fit of the form (A.5.1) is

$$D_p = 900\,(W_o U_o D_o)^{0.6} / T_o \ , \qquad \text{(A.5.2)}$$

but the scatter is such that one could just as well say $D_p \approx 0.22$ μm with maximum deviations of 0.08. The range of D_p in all the tests was 0.15–0.5 μm. These diol smokes would be excellent for marker nephelometry, and a smoke generator of this type may be useful when very large quantities of smoke are needed.

One aspect of Nolen's work is somewhat unclear. The mean boiling point of Diol 55 (mol. wt. 250–300) should be around 320–350°C. Nolen noted that only a fraction of the oil fed to the vaporizer was actually evaporated - the rest accumulated during a run. It appears from this and his operating temperatures that the vapour produced must have been composed of the lighter fractions of the diol, and the molecular weight of the vapour must have less than the mean molecular weight of the diol feed.

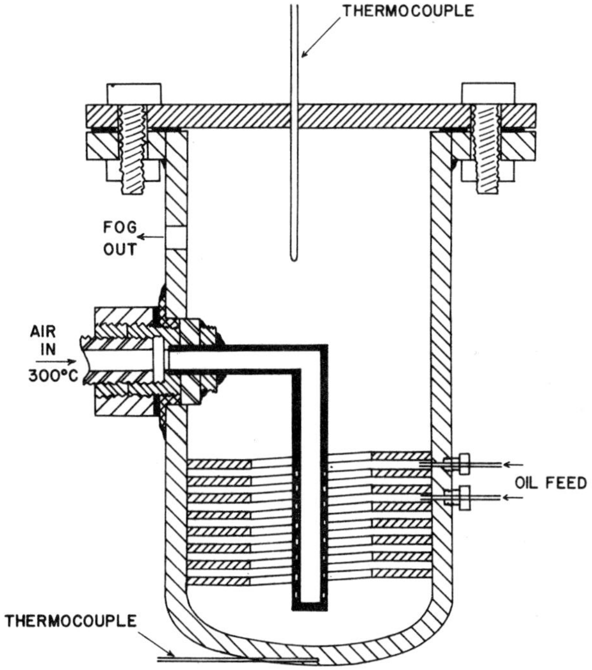

Fig. 7 The oil condensation smoke generator of Rathgeber [76] and Grandmaison [54]. The sides and top were insulated and the bottom was heated with a controlled laboratory flask heater. Oil is fed onto the spirals of the distributor surface (an aluminium helix). Air is fed into the chamber in jets from 1.5 mm dia holes in the feed pipe.

Rosensweig, Hottel and Williams [19] used diol smoke, and so have the writer and his colleagues in all their work on air/air mixing. The design of the last of several generators that have been built is shown in Fig. 7. Cold diol is injected through up to six feed lines at staggered point onto an aluminium spiral in the vaporizer chamber. The chamber is insulated and electrically heated all around, and is operated just above the boiling point of the diol. The oil runs down the spiral a short distance before completely evaporating. Hot air is blasted into the spaces of the spiral from holes in a feed pipe, and mixes thoroughly with the vapour. The product air/vapour mixture is passed through a water-cooled heat exchanger where it is quickly cooled to room temperature and the oil condensation smoke is formed. The smoke may be diluted with more air in a subsequent jet-mixing chamber.

The system has been described in detail by Rathgeber [76], with minor modifications by Grandmaison [54]. They used Diol 40, an Esso product obtained from the Imperial Oil Company of Canada. The diol was fed from an air-pressurized reservoir at 5.5 atm gauge pressure through a micro-feed valve of very high quality. The diol could be vaporized at up to 15 ml/min. The rate required in Rathgeber's work was 1 ml/min, but Grandmaison used 2–10 ml/min. The hot air was fed at a vaporizer inlet temperature of 300°C and at rates up to 0.16 kg/min. The temperature indicated by the thermocouple in the vaporizer space before the gas outlet was about 315°C. The mean particle diameter determined with a Langmuir-Schaefer cell was a little over 1 μm.

This generator produces smoke at a very steady rate over a wide range of diol feed-rates. It requires very little attention and is easy to operate. The most critical component is the oil feed valve, and it is essential that the diol be very clean to avoid irregular feeding or clogging of the valve. Diol kept for long periods should be stored in a cold room to lower the rate of formation of gum particles which eventually spoil it for use in this type of system.

A different vaporizer design, Fig.8, was employed in earlier studies. Air was passed up through a 5 cm dia by 30 cm high bed of steel rivets topped with a wad of steel wool and heated by an electric rod heater wound around the outside of the column. Oil was fed to the top of the bed and vaporized over the top 5 cm of packing. The rising air/vapour mixture was entrained by a hot air jet in a mixing chamber directly over the bed, and the resulting mixture then went to the cooler. Diol 55 – the same oil used by Nolen – was used with these vaporizers. The maximum vaporization rate was about 2 ml/min. The air/vapour temperature at exit was about 345°C. The smoke particle diameter was around 0.5 μm, rather similar to Nolen's smokes. Diol 55 boils at a higher temperature than Diol 40, and as a rule the higher the boiling temperature the smaller the particle size of a condensation smoke formed from it. These Diol 55 vaporizers gave very good service, but required maintenance every 10 h of operation, due to buildup of char residues in the bed. The vaporizer of Grandmaison and Rathgeber was designed for operation with Diol 55, hopefully with less maintenance than previous models, but then it was found that it operated better – indeed flawlessly – with Diol 40.

The difference in particle sizes between Diol 40 and Diol 55 condensation smokes might be critical in some situations, Diol 55 producing the smaller particles.

A portable oil condensation smoke generator, the Model 11-48 Cloud Maker, is manufactured by Testing Machines Inc., 72 Jericho Turnpike, Mineola, L.I., New York. The smoke output is said to be variable from a trickle to 30 m^3/min, but the operating time per charge is very short at high flow rates.

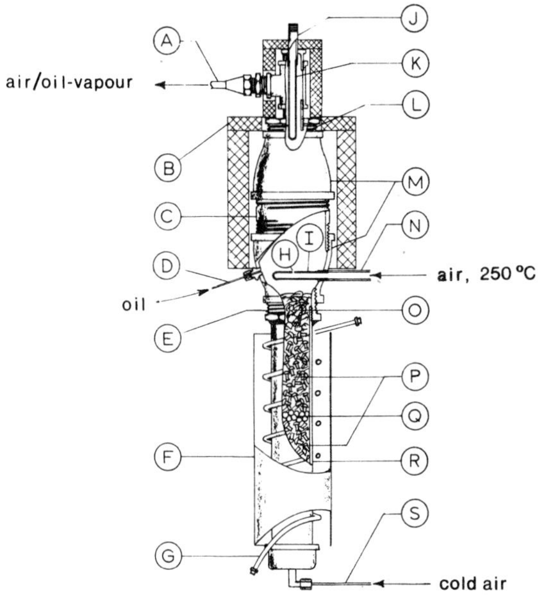

Fig. 8 The oil-condensation smoke generator of Becker, Hottel and Williams [20]. C, jet-mixing chamber of 7.6 cm dia pipe; R, vaporizer column of 3.9 cm dia by 30 cm long pipe; P, packing of steel rivets; O, plug of stainless steel wool; G, electric heater; F, thermal shield; I, air feed tube; H, 1.5 mm air inlet orifice. Oil runs down the wall into the heated packed bed where it evaporates into a small stream of air rising up the column. The resulting concentrated air/oil-vapour mixture is diluted by jet mixing with a large stream of air in the chamber above the bed.

Satisfactory smokes of liquid droplets can also be formed by air atomization. Oil micro-fog lubricators used with air-operated machinery provide oil droplets under 2 μm dia (C.A. Norgren Co., Littleton, Colorado); these devices are very cheap and should be satisfactory when rather low particle concentrations are acceptable, e.g., when only mean concentrations are to be measured.

A modified version of the collision atomizer described by Green [111] was used in marker nephelometry by Liu and Karaki [47]. The liquid was dioctyl phthalate, and particles of 2–3 μm dia were obtained. A collision atomizer and dioctyl phthalate were also used by Yang and Meroney [45,46] who adopted a design described by Van de Hulst [26]. They obtained a particle size of 4 μm.

Atomizing smoke generators are commercially available in two models from Royco Instruments, Inc., 141 Jefferson Drive, Menlo Park, California. The larger of these produces 85–510 l/min of dioctyl phthalate smoke with a particle mass concentration of 5 mg/l. The smaller one produces up to 30 l/min, and the particles are said to be 0.1–5 μm dia. Shaughnessy [23] used the larger model for marker nephelometry. He found the smoke production rate to be very steady – less than 2% total variation over a period of hours, and the mean particle size under his operating conditions was 0.3 μm. This is a very satisfactory performance.

The diol condensation smoke generator of Rathgeber and Grandmaison can produce up to 130 l/min of smoke at a concentration of over 100 mg/l, equivalent to 2600 l/min of smoke at 5 mg/l. In general, it is clear that liquid-particle smoke can be made in considerably higher concentrations by vapour condensation than by air atomization. This is not necessarily an advantage, as there are a variety of factors to be considered. At a given mass marker concentration Γ^*, the amount of light scattered when $\pi D_p > \lambda$ tends to vary inversely as the particle diameter. The number of particles at given Γ^* varies inversely with D_p^3, and this strongly affects the marker shot noise. In any case, in a given experiment there is some smoke concentration that is sufficient, and a smoke generator that delivers it is satisfactory. The use of a very much denser smoke may invite error from effects of coagulation. At the other extreme, too light a smoke heightens marker shot noise. The author and coworkers have usually diluted the product of their diol condensation smoke generators, and the diol concentration in the marked air at entry to the mixing field under study was typically 5–10 mg/l. This gives around $10^4/\text{mm}^3$ particles of 1 μm dia for the number concentration at entry.

A rather remarkable liquid particle generator, called the Model 3050 Berglund-Liu Monodisperse Aerosol Generator, is commercially produced by Thermo-Systems Inc., 2500 Cleveland Avenue North, St. Paul, Minnesota. It produces particles whose diameter is quite uniform and is given within 1% by the operating conditions of the system. Particles in the size range 0.6–15 μm dia are produced in a concentration of 200 particles/cm^3 at air flow rates up to 100 l/min. The number concentration is too low for most applications of marker nephelometry, but the system provides an excellent reference standard for characterizing particles.

In marker nephelometry on gaseous systems at elevated temperatures, solid particles of materials such as aluminium oxide, magnesium oxide, titanium dioxide, and silica must be used. Many of these materials are available as powders consisting of particles of 1 μm dia or less, from laboratory chemical suppliers and from manufacturers of pigments and abrasives. However, it is very difficult to break up agglomerates and adequately disperse the particles in a gas stream. A material called CAB-O-SIL Microgrit consists of aluminium oxide particles coated with silica, and may be suitable for use in flames. It is manufactured by Micro Abrasives Inc., 720 Southampton Road, Westfield, Massachusetts 01085. Metal oxides such as Al_2O_3 and MgO are around four times as dense as diols or dioctyl phthalate, and for this reason it is desirable that the particles be smaller in order that they have adequately low inertia to follow the fluid motion faithfully. On the other hand, gas viscosity increases with temperature, and this is a mitigating effect according to the theory in Sec.3.11.

Fluidized beds and jet mills may be considered as means for dispersing powders. The production of metal oxide aerosols in electric arcs has been described by several authors [112,113]. Gurnitz, in his study of a premixed flame by marker nephelometry [57], tried this method for making MgO smoke, but the production rate was generally much too low and unsteady to be useful.

The vapour phase hydrolysis of metal chlorides produces dense metal oxide smokes.* A simple generator for making TiO_2 smoke by contacting $TiCl_4$ (a liquid)

* Note added in proof. A footnote on p.110 gives the reference for a study of turbulent diffusion flames by Ebrahimi and Kleine. They used a TiO_2 smoke and describe a system for making it by mixing moist fuel gas with a dry fuel-gas/$TiCl_4$-vapour stream. The author is now building such a system.

with moist air, for use in flow visualization studies, is described by Maltby and Keating [109]. The oxides SiO_2 and Al_2O_3 can be made pyrogenically by introducing a stream of metal chloride vapour in dry air into the flame zone of a hydrogen flame [112]. The great drawback of these smokes is the presence of a high concentration of HCl.

Metal oxide smokes can also be prepared by atomizing water or alcohol solutions of metal salts into a flame [112] – e.g. Al_2O_3, Cr_2O_3, Fe_2O_3, TiO_2, and ZrO_2 from sulphates, Mn_3O_4 from the acetate, and ThO_2 from the nitrate. The number concentrations obtained might be too low for marker nephelometry, but adequate for laser-Doppler velocimetry.

The technique employed by Gurnitz [57] and lately by the author and coworkers has been to burn pure magnesium ribbon in a flame. This produces a dense MgO smoke, but some difficulty is experienced in maintaining a steady production rate, due to vagaries in the ribbon's burning. The magnesium ribbon is available from laboratory chemical suppliers. A generator for burning one or

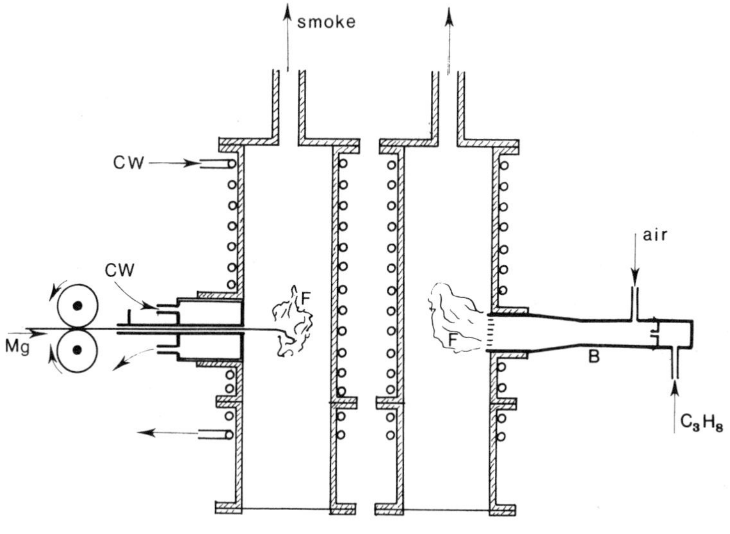

Fig.9 Magnesium oxide smoke generator. Magnesium ribbon is fed by rolls through a tube into the flame, F, of a meker burner, B, in a combustion chamber. The combustion chamber and feeder walls are cooled with cold water, CW.

two ribbons is shown in Fig.9. Gurnitz burned six ribbons in parallel. These generators need further development work; the optimum burner and feeding arrangement are yet to be found. The author is also considering the use of other metal oxides and other types of generators.

With growing interest over the last decade in air-pollution monitoring, clean rooms, air filter testing, and so forth, a variety of instrumentation has appeared for the analysis of particle size and concentration in aerosols. All that is usually necessary to characterize adequately the smokes used in marker nephelometry, though, is to determine the mass concentration by gravimetric analysis, and a mean number concentration by measurements of marker shot noise as suggested in Sec. 3.6. Sampling and filtration systems for gravimetric analysis are commercially available and inexpensive.

Safety should not be forgotten in work with smokes. The best rule to follow is that personnel should always be protected against inhaling smokes. Some smokes are obviously worse than others; one of the advantages claimed for dioctyl phthalate is that it is nontoxic.

A.6 Signal Processing

Since the signal resistor used with the phototransducer is usually of large resistance R (typically 5–50 kΩ) care must be taken that the input impedance of the first instrument to receive the signal is proportionately large, desirably at least 100 R. If necessary, an impedance transformer – a suitable cathode follower or an amplifier – can be interposed. The alternative is to calibrate the system response to make allowance for the error.

The instruments used in hotwire velocimetry and for acoustical analysis are all applicable: d.c. and random-signal voltmeters with selectable time constants of 1–100 s, spectrum analyzers, correlators, probability density analyzers, and so forth. Analogue and digital tape recorders, computers, and data acquisition systems can also be used in various ways.

A useful approach in many cases is to have a separate nephelometer probe, which can be of fairly primitive design, continually monitoring the marked feed at the input point. This is the simplest way of guarding against error due to variation in the feed smoke concentration Γ^*. Care must be taken that the scattering angle Ω_s and the incident beam radiation spectrum or operating wavelength are the same for the monitor as for the main system, otherwise the signal ratios may change with variations in the particle size or size distribution.

A ratiometer is commercially available (Model 188 Ratiometer, Princeton Applied Research Corporation, P.O. Box 2565, Princeton, New Jersey) that will take the ratio of any slowly varying signal A, to another one, B. For example, A might be the output of an rms voltmeter, representing $\hat{\gamma} \equiv \sqrt{(\overline{\gamma^2})}$, while B is the output of a d.c. voltmeter representing Γ_o; then $A/B = \hat{\gamma}/\Gamma_o$. One can also take $\overline{\Gamma}/\Gamma_o$, $\hat{\gamma}/\overline{\Gamma}$, and a variety of other values.

For very accurate work it is usually necessary to calibrate the instrument systems carefully in exactly the configurations in which they will be used (with the same signal resistor, transmission cables, grounds, etc.) and then consistently to use them in that configuration.

Spectrum analyzers are often found to show very perceptible step changes in response in switching between frequency ranges, and careful calibration will eliminate error from this source.

The effective frequency bandwidth of low-pass filters or spectral analyzers is best determined with white noise. White noise generators are available. However, electronic shot noise from a phototransducer energized by a steady light source is quite satisfactory (it should be spectrally analyzed, of course, to make sure that the spectrum is quite white in the regions to be used). Multi-speed analogue tape recorders can be used to shift white noise to a different frequency range from that in which the primary generator is effective.

The author has found it a good rule to provide wider frequency response than needed to take the interesting part of the turbulence spectrum, and then to set the cut-off frequency with a good passive lowpass filter having a sharp cut-off characteristic. A set of selectable filters with different cut-off frequencies – say 5, 10, 20, and 50 kHz – are used to insure that frequency response does not extend beyond (i) the selected cut-off frequency, or (ii) the white noise threshold.

STUDIES IN THE PREDICTION OF TURBULENT DIFFUSION FLAMES

Said Elghobashi
CHAM of North America, Inc.
Huntsville, Alabama 35810

ABSTRACT

The article concerns the problem of developing a computer-based model of turbulent diffusion flames in axisymmetric combustion chambers. Proposals are made for a simple mathematical model of chemical reaction taking account of the important effects due to the presence of fluctuations in the level of the constituent gases. A clipped-Gaussian probability density function is assumed for this purpose. The complete mathematical model entails the simultaneous solution of six transport equations describing mean and turbulence characteristics of the flow and concentration fields. A number of comparisons are drawn between experiments and predicted quantities. Generally the correct trends are displayed though in a number of respects further refinement of the reaction model appears to be needed.

1. INTRODUCTION

The present article provides an account of recent work by the author on the calculation of steady, gaseous, turbulent, diffusion flames confined in cylindrical combustion chambers. In these flames, fuel and oxidant are separately admitted into the chamber and, since the chemical reaction time-scale is often several orders of magnitude smaller than the time-scale of turbulent mixing, the mixing imposes a limit on the reaction rate thus controlling the rate of combustion. Nevertheless, chemical reaction takes place at the molecular level; molecular diffusion between fuel and oxidant must therefore take place before any reaction is possible.

Turbulent diffusion flames are common in propulsive systems and industrial furnaces. Gas-turbine combustors, power-station boilers and glass-making furnaces are some practical examples. These combustion systems employ turbulent diffusion flames because they allow a more sensitive control of the performance and eliminate the detonation hazards of premixed flames.

Recent awareness of the limits of energy resources has provided new incentive for studying these flames. To achieve higher combustion efficiencies, detailed studies of the factors that enhance the rate of turbulent mixing of fuel and oxidant are needed. Present and future fuel shortages may even force much existing combustion equipment to run with fuels other than those specified for high output and low-pollutant emission. This situation poses the problem of designing new (or modifying existing) combustion chambers to satisfy both performance and emission-control requirements. These important and difficult problems underline the urgent need for a fundamental understanding of turbulent diffusion flames in practical combustion systems.

The currently available experimental evidence (Hawthorne *et al.* [1]; Pompei and Heywood [2]; Bilger and Kent [3]) suggests that the instantaneous concentration of a species at a point in a turbulent flame experiences large random fluctuations. Because of these concentration fluctuations we find that, even in the fast-chemistry limit discussed by Professor Libby, the time-mean mass fractions of both reactants are non-zero at the same location. This "unmixedness" is responsible for the appearance of the brush-like thick reaction zones of turbulent diffusion flames. Thus, a complete characterization of these flames must involve specification in the flow field of both the time-mean values and the variation with time of properties such as concentrations, temperature, density and velocity. Accurate predictions of their behaviour should take account of some, if not all, of the effects of these fluctuations on the time-mean values of the flow properties.

In the theoretical study of turbulent diffusion flames one encounters two main problems, one mathematical, the other physical. The mathematical problem, which is common to all turbulent flows, reacting and non-reacting, arises because the rigorous equations governing the instantaneous structure of the flow are coupled, non-linear, partial differential equations (see, for example, Favre [4], Bray [5]). In flames, the magnitude of the problem increases because the number of dependent variables is larger than that for a non-reacting flow (density, enthalpy, concentration, reaction rates) and hence the number of equations is greater. Although numerical procedures exist for solving these equations, the computational mesh required to cover the smallest eddy sizes is too fine to be handled by present-day computers. The usual way of avoiding this impasse is to decompose the terms of the instantaneous equations into their time-mean and fluctuating parts, resulting in a set of equations with more unknowns than equations [5]. In order to close the set of equations a "turbulence model" is needed to approximate the fluctuating correlations in terms of known time-mean quantities. The physical problem arises from our inadequate knowledge of the laws that govern both turbulence transport and the chemistry of reaction. This lack of knowledge makes it difficult to construct satisfactory turbulence and reaction models.

The writer [6] has surveyed the various theoretical approaches to the calculation of turbulent diffusion flames and concluded that:

(a) Numerical procedures which incorporate the effects of turbulent concentration fluctuations on the rates of chemical reaction appear to offer the only reliable approach to predicting turbulent flames (e.g., Spalding [7]; Kent and Bilger [8]).

(b) The effects of combustion on the flow through the spatial variation of the time-mean density should take account of the influences of concentration and temperature fluctuations on the density (see, for example, Kent and Bilger [8]).

So far as is known, no previous work has incorporated both the above effects in a calculation procedure for turbulent diffusion flames. The present article reports the writer's contribution to the development of such a scheme. Section 2 of the article describes the mathematical and physical model adopted, Section 3 gives an outline of the method of numerical solution while in Section 3 we draw comparisons between the computed behaviour of various confined flames and available experiments.

2. THE MATHEMATICAL AND PHYSICAL MODEL

2.1 *Background*

Here we introduce three sets of equations that are proposed as descriptions of (i) the effective momentum, heat and species fluxes due to turbulence, (ii) chemical reaction in a turbulent flow and (iii) concentration fluctuations in turbulent diffusion flames. The analysis is restricted to steady, axially symmetrical situations where recirculation may occur and where all constituents are gaseous. This system of partial differential equations provides, together with the mean-flow conservation equations, a closed set for the prediction of both the fluid dynamic and thermodynamic properties of the flow. The equation set characterizing the flame model is developed in the next section. The mean-flow equations are presented first then the various physical models are cast into mathematical form. Finally, the numerical procedure used to solve the equation set is briefly described in Sec.2.3.

2.2 *Equations and Models for Turbulent Reacting Flows*

2.2–1 The time-mean motion

The time-mean motion of a turbulent flow is governed by the laws of mass and momentum conservation. For axisymmetric flows these laws can be expressed in terms of a stream function ψ and a vorticity ω (Wolfshtein [9], Gosman *et al.* [10]) defined as follows:

$$\psi \equiv \int [\rho r(udr - vdz)] \quad , \tag{2.1}$$

$$\omega \equiv \frac{\partial v}{\partial z} - \frac{\partial u}{\partial r} \tag{2.2}$$

where u and v are the time-mean velocity components in the axial and radial directions respectively. The two resulting equations are non-linear, partial differential equations of elliptic type; they can be cast in the following common form, in cylindrical-polar coordinates:

$$a\left(\frac{\partial}{\partial z}\left(\phi\,\frac{\partial\psi}{\partial r}\right)-\frac{\partial}{\partial r}\left(\phi\,\frac{\partial\psi}{\partial z}\right)\right)-\frac{\partial}{\partial z}\left(br\,\frac{\partial(c\phi)}{\partial z}\right)-\frac{\partial}{\partial r}\left(br\,\frac{\partial(c\phi)}{\partial r}\right)+rd=0 \qquad (2.3)$$

Convection terms — *Diffusion terms* — *Source term*

In the above equation, ϕ stands for either ψ or ω/r, and the meanings of the functions a, b, c and d are given below in Table 2.1.

Table 2.1 **The significance of the coefficients in the stream function and vorticity equations.**

ϕ	a	b	c	d
ψ	0	$1/(\rho r^2)$	1	$-\omega/r$
ω/r	r^2	r^2	μ_{eff}	$r\left(\frac{\partial}{\partial r}\frac{(u^2+v^2)}{2}\frac{\partial\rho}{\partial z}-\frac{\partial}{\partial z}\frac{(u^2+v^2)}{2}\frac{\partial\rho}{\partial r}\right)$

The use of these dependent variables avoids the need to consider directly the pressure and the two component velocities, u, v. When needed, the velocities can be recovered from the stream function via the definition (2.1) and the pressure from the r- and z-direction momentum equations [10].

From Table 2.1, it is clear that in order to solve the equations for vorticity and stream function it is necessary to provide information about ρ, the time-mean density, and about μ_{eff}, the effective viscosity. First, we consider how to approximate μ_{eff}.

2.2–2 The turbulence model

As remarked in the Introduction, the exact time-dependent conservation equations of mass, momentum, concentration, etc., cannot be solved for practically interesting turbulent flows because:

(i) the equations are non-linear and of partial-differential type;

(ii) turbulence is a three-dimensional motion, and many of the important processes take place on very small scales. Therefore, numerical techniques which calculate the dependent variables at discrete points in space would need approximately 10^6 grid points to cover even one cubic centimetre of space.

Fortunately, experience with turbulent flows in pipes, boundary layers on walls and free jets suggests that there is no need to focus on time-dependent character of turbulence; for the majority of circumstances the time-averaged equations provide sufficient information about the flow. Also there is no need for excessively fine grids, since the time-averaged properties vary with position far more gradually than the instantaneous flow properties. In the process of time-averaging, however, correlations of fluctuating velocity appear in the mean momentum equations as unknowns. There has recently been a number of reviews of methods of approximating these correlations (Launder and Spalding [47], Mellor and Herring [11] and Reynolds [12]). No attempt will be made here to duplicate these reviews. The writer's work has used exclusively the so-called '$k \sim \epsilon$ viscosity' model in which the effective turbulent viscosity, μ_t, is related to two scalar properties of turbulence: its kinetic energy per unit mass, k and its rate of dissipation of that energy, ϵ. The connecting formula is:

$$\mu_t = c_\mu \rho k^2 / \epsilon \tag{2.4}$$

where c_μ is usually taken as a constant. The effective viscosity appearing in Table 2.1 is then just the sum of the molecular and turbulent contributions, i.e.

$$\mu_{eff} = \mu + \mu_t \; . \tag{2.5}$$

The quantities k and ϵ are obtained from the solution of two differential equations as described shortly.

Turbulence models of the above kind have a number of attractive features. They are tolerably simple and thus do not increase unreasonably the cost of computations over that for the corresponding laminar flow. They also possess significantly greater width of applicability than do simpler closures of the one-equation or zero-equation type (see Launder *et al.* [13] for a series of comparative predictions with different types of closure). Moreover, what appears to be their weakest feature – the transport equation for ϵ – is shared also by more elaborate turbulence models which find the effective stress correlations from transport equations. Of course, it is by no means certain that the model will prove generally adequate in predicting flames; as Professor Libby has discussed in the first article of this volume, the presence of large density fluctuations provides another mechanism of momentum transfer. The approach we have followed is, nevertheless, to start with the form of the model that many writers have employed successfully for non-reacting flows and to see what level of predictive accuracy is achieved in calculating the characteristics of turbulent flames.

Formally, the kinetic energy of turbulence k is defined as:

$$k \equiv \langle u_i' u_i' \rangle / 2 \tag{2.6}$$

where u_i''s are the components of velocity fluctuations in the three space directions and the convention whereby repeated indices are summed is adopted. The transport equation for k has been considered by many authors (e.g.,

Hinze [14], Bray [5]). Gosman *et al.* [10] introduced a modelled k equation in cylindrical coordinates cast in the general form of Eq. (2.3). This form is adopted here; the meaning of the corresponding functions a, b, c and d is given in Table 2.2.

Table 2.2 The form of the turbulence-energy and dissipation rate transport equations.

ϕ	a	b	c	d
k	1	μ_{eff}/σ_k	1	$\rho\epsilon - G_k$
ϵ	1	$\mu_{eff}/\sigma_\epsilon$	1	$c_2\rho\epsilon^2/k - c_1 G_k \epsilon/k$

The generation expression for k, G_k is given by:

$$G_k \equiv \mu_t\{2[(\partial u/\partial z)^2 + (\partial v/\partial r)^2 + (v/r)^2] + (\partial u/\partial r + \partial v/\partial z)^2\} .$$

The rate of viscous dissipation of turbulent kinetic energy at large turbulence Reynolds numbers may be defined as:

$$\epsilon \equiv \nu\overline{\langle(\frac{\partial u_i}{\partial x_j})^2\rangle} . \tag{2.7}$$

In the definition (2.7), the x_j's are the distances along the coordinates r, z and θ; ν is the kinematic viscosity, and the summation convention is used. Davydov [15] and Harlow and Nakayama [16] derived the exact transport equation for ϵ for a uniform-property flow. The modelled form shown in Table 2.2 is that which has evolved from the work of Daly and Harlow [17], Hanjalić and Launder [18] and Jones and Launder [19]. The form given is that appropriate to high-Reynolds-number axisymmetric recirculating flows without swirl. The values assigned to all the constant coefficients in the turbulence model appear in Table 3.2 on page 160.

It is not easy, from a knowledge of k and ϵ alone, to acquire a 'feeling' for the structure of the turbulence. It is, however, possible to deduce from k and ϵ some characteristic length and time scales that can be easily compared with characteristic scales of the flow. A local length scale of turbulence, ℓ, proportional to the size of energy-containing eddies can be defined as:

$$\ell \equiv k^{3/2}/\epsilon ; \tag{2.8}$$

while the corresponding time-scale, t, is obtained from:

$$t \equiv k/\epsilon \tag{2.9}$$

The time-mean density. Equation (2.4) indicates that μ_{eff} can be obtained only if ρ, the time-mean density, is known at every point in the flow field; a knowledge of ρ is also in the stream-function and vorticity equations. In a turbulent reacting flow, the instantaneous density at any point exhibits large fluctuations caused by the fluctuations of temperature and concentration (Kent and Bilger [8], Bray [5]). The instantaneous mixture density can be written as:

$$\tilde{\rho} = \frac{\tilde{p}}{\mathcal{R}} \Big/ \tilde{T} \sum_j \left(\frac{\tilde{m}_j}{M_j}\right) \quad , \tag{2.10}$$

where $\tilde{\rho}$, $\tilde{p}$, $\tilde{T}$ and $\tilde{m}_j$ stand for the instantaneous values of density, absolute pressure, temperature and the j-species mass-fraction respectively; $\mathcal{R}$ is the universal gas constant and M_j is the molecular weight of the j-species. Defining $\tilde{a}$ as $\sum_j \left(\frac{\tilde{m}_j}{M_j}\right)$, and writing: $\tilde{\rho} \equiv \rho + \rho'$, $\tilde{T} \equiv T + T'$, $p \equiv \tilde{p} + p'$ and $\tilde{a} \equiv a + a'$,

we can express the time-mean density as:

$$\rho = \frac{p}{\mathcal{R}aT} \left[1 - \frac{\langle T'a' \rangle}{Ta} + \frac{\langle T'a' \rangle^2}{T^2 a^2} - \ldots\ldots\right] . \tag{2.11}$$

It can be seen that neglecting the first and second order corrections in Eq. (2.11) will over-estimate ρ in regions where $\langle T'a' \rangle / Ta$ is not much smaller than unity. Thus, to evaluate correctly the time-mean density, it is necessary to include the effects of the fluctuations in both temperature and concentration. Two models are required: the first is for the chemical reaction whereby the instantaneous values of temperature and concentration are obtained; the second is a model for the fluctuations of concentration which provides a means of evaluating the time-mean properties. Such models will be discussed in turn in the following sections.

2.2–3 The chemical-reaction model

The model proposed here is the fast-chemistry idealisation discussed on pp.30 *et seq.* by Professor Libby. The central assumption is that, whenever fuel and oxidant both exist at a point, chemical reaction will instantaneously proceed to completion in a single step, producing combustion products. This hypothesis of instantaneous chemical equilibrium is not far from reality in turbulent-un-premixed flames because the time of recombination of the non-equilibrium radicals in hydrogen-rich flames is of the order of 10^{-6} s, while the smallest time-scale of turbulence is typically about 10^{-4} s (Günther and Simon [20]).

From the above assumptions and from the work of Shvab [21], Zeldovich [22], Williams [23] and Gosman *et al.* [10], it may be concluded that a suitably chosen linear combination of the equations describing the conservation of two un-premixed reactants (a fuel and an oxidant) yields an equation whose form is identical to that describing the convection and diffusion of chemically inert species. The equation which results will have no source term, provided that:

(a) the turbulent transport coefficients for the two reactants are equal at every point in the field, and
(b) 1 kg of fuel always combines with (i) kg of oxidant to produce ($1+i$) kg of combustion products.

The dependent variable, the mixture fraction f, of the resultant equation is defined as:

$$f \equiv (\gamma - \gamma_o)/(\gamma_1 - \gamma_o) \ , \tag{2.12}$$

where γ is given by

$$\gamma \equiv m_{fu} - m_{ox}/i \ , \tag{2.13}$$

or, equivalently,

$$\gamma \equiv m_{fu} + m_{pr}/(1+i) \ , \tag{2.14}$$

where i is the mass of oxidant required for the complete combustion of a unit mass of fuel, and the subscripts 1 and 0 denote the initial conditions of the un-premixed fuel and oxidant streams respectively. The conservation equation for f is a partial differential one of elliptic type (Gosman *et al.* [10]) and can again be written in the general form of Eq. (2.3); the corresponding meanings of a, b, c and d are given below in Table 2.3.

Table 2.3 The form of the mixture fraction equation.

ϕ	a	b	c	d
f	1	μ_{eff}/σ_f	1	0

Since we need instantaneous values of the mass fractions of the reactants, the definitions given by (2.12–14) can be written in an instantaneous form, using the following relations:

$$\tilde{f} \equiv f + f' \ , \tag{2.15}$$

$$\tilde{\gamma} \equiv \gamma + \gamma' \ , \tag{2.16}$$

$$\tilde{m}_j \equiv m_j + m_j' \ , \tag{2.17}$$

The basic assumption of instantaneous chemical equilibrium is used to obtain the important relations:

$$\tilde{m}_{fu} = 0, \quad \tilde{m}_{ox} = m_{ox,0}(1 - \tilde{f}/f_{st}) \quad 0 \leqslant \tilde{f} \leqslant f_{st} \ : \tag{2.18}$$

$$\tilde{m}_{ox} = 0, \quad \tilde{m}_{fu} = m_{fu,1}(\tilde{f} - f_{st})/(1 - f_{st}) \quad f_{st} \leqslant \tilde{f} \leqslant 1 \ : \tag{2.19}$$

where f_{st} is the stoichiometric value of the mixture fraction, being obtained from (2.12) and (2.13) as:

$$f_{st} = 1/[1 + i(m_{fu,1} / m_{ox,0})] \quad . \tag{2.20}$$

The mass fraction of the products of combustion is obtained from:

$$\tilde{m}_{pr} = 1 - \tilde{m}_{fu} - \tilde{m}_{ox} \tag{2.21}$$

but their composition (including whatever inert gases are present) is obtained from stoichiometric considerations as will be seen later.

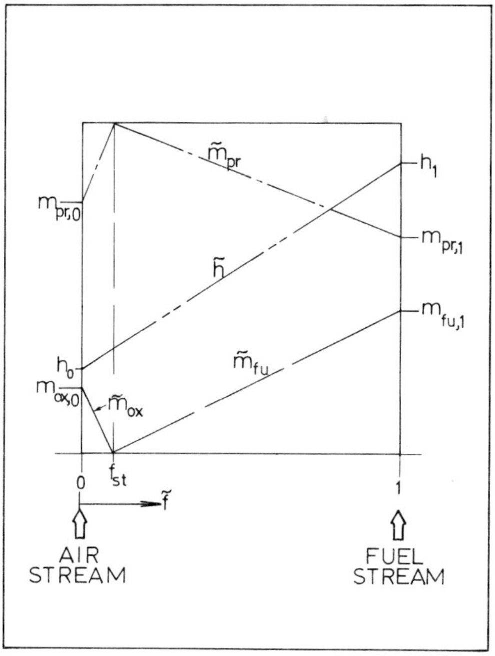

Fig. 1 Dependence of mass concentrations and enthalpy on the mixture fraction.

Figure 1 illustrates the linear relationship between the instantaneous mass fractions of the reactants and $\tilde{f}$ as expressed by Eqs (2.18–21). The linear variation of the enthalpy $\tilde{h}$ with $\tilde{f}$ requires that, in addition to conditions (a) and (b) above, the three following constraints should be satisfied [22, 10].

(c) the boundaries of the combustion system are adiabatic;

(d) the Lewis number (the ratio of thermal diffusivity to the diffusion coefficient of concentration) is unity, and

(e) both the kinetic energy of the mean flow and of turbulence are negligible compared with the heat of reaction H.

The relationship between h and f can be written:

$$f \equiv (h - h_o)/(h_1 - h_o) \quad , \tag{2.22}$$

where

$$h = \int_0^T c_p dT + Hm_{fu} \ . \tag{2.23}$$

In Eq. (2.23), c_p is the specific heat at constant pressure, as in (2.12), subscripts 0 and 1 denote the initial conditions of the un-premixed oxidant and fuel streams respectively.

Ref. [10] has derived a conservation equation for h which is similar in form to Eq. (2.3) with a zero source term d as in the f equation. The five conditions (a–e) above, though advantageous in reducing by two the total number of equations to be solved, do not allow the inclusion of heat transfer to or from the system.

The linear variation of $\tilde{h}$ with $\tilde{f}$ in Fig.1 is obtained from (2.22) by writing $\tilde{h}$ as

$$\tilde{h} \equiv h + h' \ ; \tag{2.24}$$

$\langle h' \rangle = \langle f' \rangle = 0$ for stationary turbulence.

It follows that:

$$\tilde{h} = h_o + \tilde{f}(h_1 - h_o) \ , \tag{2.25}$$

$$\tilde{T} = (\tilde{h} - H\tilde{m}_{fu})/\sum_j \tilde{m}_j c_{p_j} \ , \tag{2.26}$$

where

$$h_o = c_{p\,ox,o}\, T_o + c_{p\,inert,o}\, T_o \ , \tag{2.27}$$

and

$$h_1 = c_{p\,fu,1}\, T_1 + c_{p\,inert,1}\, T_1 + Hm_{fu,1} \ . \tag{2.28}$$

To evalute $\tilde{T}$ from (2.26) the summation term, $\sum_j \tilde{m}_j c_{p_j}$, is calculated from:

$$\sum_j \tilde{m}_j c_{p_j} = \tilde{m}_{ox} c_{p_{ox}} + \tilde{m}_{fu}\, c_{p_{fu}} + \tilde{m}_{pr}\, c_{p_{pr}} \ . \tag{2.29}$$

The specific heats, $c_{p_{ox}}$, $c_{p_{fu}}$ and $c_{p_{pr}}$, at a given point in the flow, are taken as functions of the chemical compositions of the oxidant, fuel and combustion products respectively. Moreover, it is assumed that the specific heats follow a third-order, polynomial-type variation with the local time-mean temperature (Glasstone [24], Tribus and Desoto [25]). The composition of the combustion products is obtained from the stoichiometric reaction of a hydrocarbon fuel, C_nH_{2n+2}, diluted with inert and air:

$$C_nH_{2n+2} + \beta_2 CO_2 + \beta_3 CO + \beta_1 N_2 + a_1 O_2 + 3.76 a_1 N_2 \longrightarrow$$

$$a_3 H_2O + (3.76 a_1 + \beta_1) N_2 + (a_2 + \beta_2) CO_2 \tag{2.30}$$

where the α_i's and β_i's stand for the special mol-fraction. To solve (2.30) for the composition of the products of reaction, three algebraic atom-balance equations are written for the carbon, hydrogen and oxygen atoms. The three equations yield the values of the three unknowns α_1, α_2 and α_3.

Knowledge of the instantaneous values of the mass fractions and temperature gives the value of the instantaneous density ρ via the perfect gas law:

$$\tilde{\rho} = p/\mathcal{R}\tilde{T} \sum_j (\tilde{m}_j/M_j) \ . \tag{2.31}$$

What remains now is the calculation of the *mean* density; as remarked earlier, this requires a model for concentration fluctuations.

2.2–4 The model for concentration fluctuations

Background. In order to obtain the time-mean value, θ, of a statistically-steady, randomly fluctuating quantity $\tilde{\theta}(t)$, varying with time as shown in Fig.2,

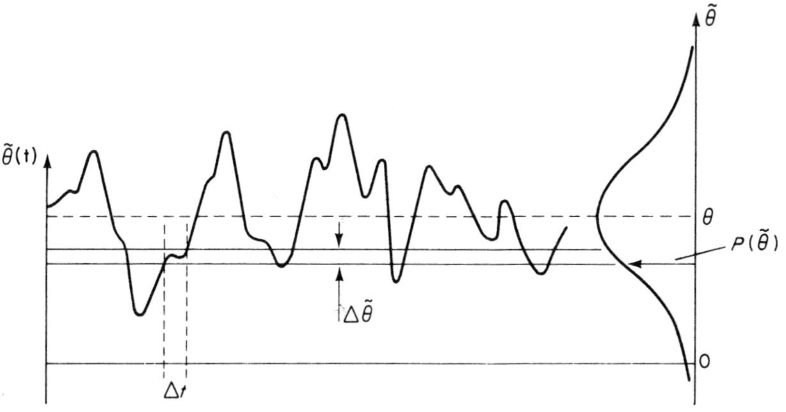

Fig.2 Typical variation of $\tilde{\theta}(t)$ and $P(\tilde{\theta})$.

one needs information about its probability density function $P(\tilde{\theta})$ which may be defined through:

$$P(\tilde{\theta})\,\Delta\tilde{\theta} \equiv \lim_{t \to \infty} \frac{1}{t}\,\Sigma(\Delta t) \ . \tag{2.32}$$

The definition (2.32) implies that $P(\tilde{\theta})\Delta\tilde{\theta}$ is the fraction of time that $\tilde{\theta}(t)$ spend between $\tilde{\theta}$ and $\tilde{\theta} + \Delta\tilde{\theta}$, and that over a long time interval the sum of the values of $P(\tilde{\theta})\Delta\tilde{\theta}$ must approach unity, i.e.

$$\int_{-\infty}^{+\infty} P(\tilde{\theta})\, d\tilde{\theta} = 1 \ . \tag{2.33}$$

From a knowledge of $P(\tilde{\theta})$, the time-mean value θ can be obtained from the definition:

$$\theta \equiv \lim_{t \to \infty} \frac{1}{t} \int_{t_0}^{t_0+t} \tilde{\theta}(t)\, dt \ . \tag{2.34}$$

From Eqs (2.32) and (2.34) we get:

$$\theta = \int_{-\infty}^{+\infty} \tilde{\theta} P(\tilde{\theta})\, d(\tilde{\theta}) \ , \tag{2.35}$$

which is simply the first moment of $P(\tilde{\theta})$ about the origin, $\tilde{\theta} = 0$.

For some other time-dependent variable $\tilde{a}$ which is a function of $\tilde{\theta}$, the time-mean value, a, can be obtained from:

$$a = \int_{-\infty}^{+\infty} \tilde{a}(\tilde{\theta})\, P(\tilde{\theta})\, d\tilde{\theta} \ . \tag{2.36}$$

Equation (2.36) forms the basis of the time-averaging process used here for all variables such as density, temperature and mass fractions which are dependent on the mixture fraction $\tilde{f}$ only. For example, if one requires the time-mean mass fraction of fuel, m_{fu}, we have to obtain first $P(\tilde{f})$, the probability density function for $\tilde{f}$; then with the help of (2.19) for $\tilde{m}_{fu}$, (2.36) leads us to the value of m_{fu}.

Unfortunately, no measurements have been reported, so far as the writer is aware, which enables us to determine $P(\tilde{f})$ for turbulent diffusion flames. Until such measurements become available, we have to assume a plausible, though necessarily somewhat arbitrary form for $P(\tilde{f})$. Two parameters are required to specify the probability density function. The first is the mean-value f; the other is the mean-square departure from the mean-value or the variance, g, which is defined as:

$$g \equiv [\tilde{f} - f]^2 \equiv \langle f'^2 \rangle \ . \tag{2.37}$$

We first discuss how g can be calculated; afterwards, a proposal for $P(\tilde{f})$ is made.

The equation for g. The transport equation for the mean-square concentration fluctuations in isothermal flows has been derived by several authors, e.g., Csanady [26], Spalding [27]. Bray [5] derived a rigorous time-mean equation for the square of concentration fluctuations for chemically-reacting flows. Independently, Csanady [26] and Spalding [7] approximated the g-equation for isothermal flow in a form similar to that of the k equation for boundary-layer flows. Spalding [7] used this modelled g equation in conjunction with a 2-equation viscosity model to predict free turbulent diffusion flames. The present study adopts essentially Spalding's model. Because recirculating flows are the ones of interest, however, we must include in the generation and diffusion processes terms containing the gradients of f and g, in both the r- and z-directions. The final equation for g can be written in the same form as (2.3); the source term, d, adopted is given in Table 2.4.

Table 2.4 The coefficients in the g-transport equation.

ϕ	a	b	c	d
g	1	μ_{eff}/σ_g	1	$c_{g2}\,\rho\epsilon g/k - c_{g1}\,\mu_{eff}\,[(\partial f/\partial z)^2 + (\partial f/\partial r)^2]$

It should perhaps be repeated that in the present work it is assumed that chemical reaction exerts its influence on g (as also on f, k, ϵ, ψ and ω) only through the time-mean density ρ in the convection and diffusion terms. Fluctuations of the rate of reaction may also influence directly the time decay rate of concentration fluctuations. This analysis will not be pursued here, though an investigation of this problem is required.

The probability-density function $P(\tilde{f})$. A satisfactory choice of $P(\tilde{f})$ must conform at least with the rudimentary physics of concentration fluctuations in turbulent flames; that is, $\tilde{f}$ should vary randomly with time and should be bounded between 0 and 1 (from the definitions expressed by (2.12–17)).

Different assumptions have been adopted in the past by various authors; some used random wave forms and others chose deterministic ones. Hawthorne *et al.* [1] and Thring and Newby [28] used a Gaussian distribution extending from $-\infty$ to $+\infty$. Richardson *et al.* [29] employed a β-function distribution for values of the mixture fraction between 0 and 1 without taking account of the fraction of time spent by the concentrations at those limits. Bilger and Kent [3] used a symmetrically truncated Gaussian distribution centred about the time-mean mixture fraction value. The truncation limits were ± 2 which allowed the instantaneous mass fractions to assume values less than 0 or greater than 1. The intermittency level was then used as a weighting factor to ensure non-negative time-mean mass fractions. Spalding [28] used a probability density function corresponding to a sinusoidal distribution of $\tilde{f}$ with time for an isothermal free jet and a battlement-shape wave for a free-jet turbulent diffusion flame (Spalding [7]). Rhodes and Harsha [30] employed a triangular probability density function for computational simplicity, but they commented that a β-function distribution would have been more realistic. Bush and Fendell [31] suggested a triangular-wave form for $\tilde{f}$, bounded between $\tilde{f} = 0$ and 1. The corresponding probability density is represented by two delta functions at the boundary limits and uniform density in between.

Naguib [32] used a clipped* Gaussian distribution between $\tilde{f} = 0$ and 1 to predict the intermittency in isothermal and reacting free jets.

In the present study, $P(\tilde{f})$ is also taken to be of the clipped Gaussian type, as shown in Fig.3. In terms of $\tilde{f}$, $P(\tilde{f})$ can be expressed as:

---

* The meanings of truncated and clipped Gaussian distributions are given by Bendat and Piersol [33] among others.

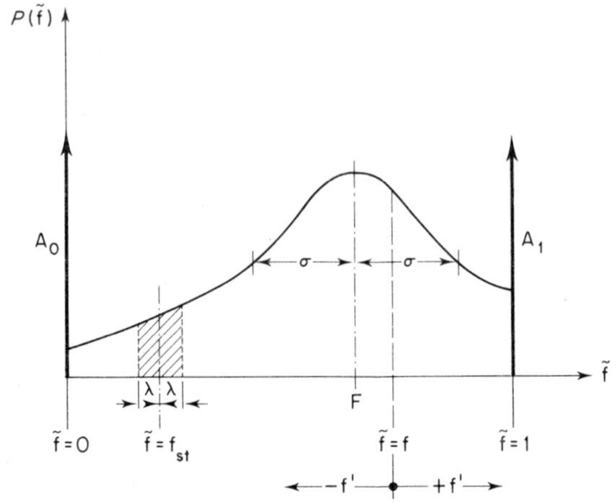

Fig.3 The clipped Gaussian distribution.

$$P(\tilde{f}) = e^{-(\tilde{f}-F)/2\sigma^2}\,[U(\tilde{f}-0) - U(\tilde{f}-1)]\,/\,\sigma\sqrt{2\pi}$$

$$+\,2A_o\,\delta(\tilde{f}-0) + 2A_1\,\delta(\tilde{f}-1)\ , \qquad (2.38)$$

where $U(\xi)$ is a unit step function defined by $U(\xi) = 0,\ \xi < 0;\ U(\xi) = 1,$ $\xi > 0$ and $\delta(\xi)$ denotes the Dirac delta function defined by: $\int_a^b y(\xi)\,\delta(\xi - x)\,d\xi = \frac{1}{2}y(x)$, if $x = a$ or $x = b$. The quantities A_o and A_1 are given by:

$$A_o = \int_{-\infty}^{0} e^{-(\tilde{f}-F)^2/2\sigma^2}\,d\tilde{f}\Big/(\sigma\sqrt{2\pi})\ , \qquad (2.39)$$

$$A_1 = \int_{1}^{\infty} e^{-(\tilde{f}-F)^2/2\sigma^2}\,d\tilde{f}\Big/(\sigma\sqrt{2\pi})\ . \qquad (2.40)$$

F and σ are respectively the mean and standard deviation for the complete Gaussian distribution, a portion of which forms part of $P(\tilde{f})$, Fig.3. The time mean value, f, and the variance, g, of $P(\tilde{f})$ are respectively given by the first moment about $\tilde{f} = 0$ and the second moment about the mean value of the area under $P(\tilde{f})$:

$$f = A_1 + \int_0^1 \tilde{f}\,e^{-(\tilde{f}-F)^2/2\sigma^2}\,d\tilde{f}\Big/(\sigma\sqrt{2\pi})\ , \qquad (2.41)$$

$$g = A_1 - f^2 + \int_0^1 \tilde{f}^2\,e^{-(\tilde{f}-F)^2/2\sigma^2}\,d\tilde{f}\Big/(\sigma\sqrt{2\pi})\ . \qquad (2.42)$$

Thus, at any given point, when both f and g are known from the solution of the differential equation set of the type (2.3), solution of (2.41) and (2.42) will yield the corresponding values for σ and F, and hence $P(\tilde{f})$ via Eq. (2.38). The solution of (2.41) and (2.42) can be speedily accomplished with the aid of a table of values of σ and F against corresponding values of f and g. It is interesting to note that F may assume values less than zero or greater than unity while f can not. We shall examine further the implied distributions of $P(\tilde{f})$ later in conjunction with predictions of g.

Evaluation of the time-mean properties. The time-mean mass fraction of fuel, m_{fu}, is given by:

$$m_{fu} = m_{fu,1} A_1 + \frac{1}{\sigma\sqrt{2\pi}} \int_{f_{st}}^{1} [(\tilde{f} - f_{st})/(1 - f_{st})] \, e^{-(\tilde{f} - F)^2/2\sigma^2} \, d\tilde{f} . \quad (2.43)$$

Likewise the time-mean mass fraction of oxidant is obtained from:

$$m_{ox} = m_{ox,0} A_o + \frac{1}{\sigma\sqrt{2\pi}} \int_0^{f_{st}} [1 - \tilde{f}/f_{st}] \, e^{-(\tilde{f} - F)^2/2\sigma^2} \, d\tilde{f} . \quad (2.44)$$

The time-mean mass fraction of combustion products, m_{pr}, is then found from:

$$m_{pr} = 1 - m_{ox} - m_{fu} \quad . \quad (2.45)$$

Similarly we obtain the time-mean temperature, T, as:

$$T = A_o T_o + A_1 T_1 + \frac{1}{\sigma\sqrt{2\pi}} \int_0^1 [(\tilde{h} - \tilde{m}_{fu} H)/\sum_j \tilde{m}_j c_{p_j}] \, e^{-(\tilde{f} - F)^2/2\sigma^2} \, d\tilde{f} \quad (2.46)$$

and the time-mean density, ρ:

$$\rho = A_o \rho_o + A_1 \rho_1 + \frac{p}{\Re\sigma\sqrt{2\pi}} \int_0^1 [1/\tilde{T} \sum_j (\tilde{m}_j/M_j)] \, e^{-(\tilde{f} - F)^2/2\sigma^2} \, d\tilde{f} . \quad (2.47)$$

In deriving Eq. (2.47), the simplifying assumption has been introduced that the joint probability density for temperature and composition fluctuations is the same as $P(\tilde{f})$.

In turbulent flames, a complete picture of the temperature field requires knowledge of both the time-mean value and the root mean square value of temperature fluctuations, $<T'^2>^{1/2}$. The latter quantity can be obtained as follows:

$$<T'^2> \equiv <[\tilde{T} - T]^2> \equiv <(\tilde{T})^2> - T^2 \quad (2.48)$$

and from

$$<T'^2> = \frac{1}{\sigma\sqrt{2\pi}} \int_0^1 (\tilde{T})^2 \, e^{-(\tilde{f} - F)^2/2\sigma^2} \, d\tilde{f} + A_o T_o^2 + A_1 T_1^2 - T^2 . \quad (2.49)$$

In the same way, we can obtain the mean-square of the fluctuations of any of the mass fractions (fuel, oxidant or products), $<m_{fu}'^2>$, $<m_{ox}'^2>$, etc..

The calculation of the time-mean density ρ from Eq. (2.38) makes it possible now to close the equation set (2.3) for the six dependent variables, ψ, ω/r, f, k, ϵ and g. The following section describes briefly how the equations are solved.

2.3 *Solution of the Equations*

The finite-difference procedure. The finite-difference procedure, used here to solve the system of equations of type (2.3), is basically that described by [10]. The fundamentals of the method are discussed in detail in that reference and are not repeated here. Only the features that have been slightly modified are now mentioned briefly.

(a) The flow domain is divided into a number of rectangular, finite-difference cells. These are arranged so that the boundaries of the flow coincide with the appropriate edges of the cells along the boundaries. Associated with each cell is a node. The relative positions of the cell walls and cell nodes are such that the walls are always situated mid-way between two adjacent nodes.

(b) By integrating the differential equations over the cell volume, finite-difference equations are derived which connect the value of a variable at a node to the values of the variable at the neighbouring nodes. In deriving these equations, two main principles are observed. First, the convective flux through a cell wall is always equal to the mass flow rate through the wall multiplied by the value of the entity concerned prevailing at the adjacent node *upstream* of the cell wall; secondly, diffusive transport leaving through one side of a cell boundary exactly equals that entering the adjacent cell through the same wall. The former practice is responsible for the stability of the numerical procedure at high Reynolds numbers, and the latter helps to ensure the satisfaction of the conservation principle over the flow domain.

(c) For variables like f, g, k and ϵ the difference equations for all continuous columns of cells between two boundary lines are solved simultaneously using a tri-diagonal matrix algorithm. Because of the strong inter-linkages between ψ and ω, however, through the no-slip condition at the wall, their difference equations are solved simultaneously by a two-variable algorithm.

(d) Starting from one end, all the variables along a line of cells are updated in turn before the next line of cells is swept. The sweep continues until the whole flow domain is covered. During a sweep all the coefficients in the difference equations are kept unchanged; they are updated before another sweep of the field is made. The field is repeatedly swept until a converged solution is obtained.

There are two main advantages that the above modifications bring over the practices proposed in [10].

(i) The conservation laws are now satisfied throughout the whole flow field. In the original procedure half-cells adjacent to the flow boundaries and the axis of symmetry were omitted from the appreciable overall budget; this omission occasionally led to significant error in the global conservation of the dependent variables.

(ii) Significant savings in computer time are realised by using a line-by-line instead of a point-by-point solution of the difference equations.

Special considerations near a wall. The physical model of turbulence transport discussed in Sec.2.2 is valid only for fully turbulent flows, i.e. where $\mu_t/\mu >> 1$. Close to solid walls, there are regions where viscous transport exceeds that due to turbulence. An economical method of accounting for these regions is through the use of wall-functions. These consist, in essence, of algebraic formulae between the flow-variables in the fully-turbulent region and the associated fluxes at the wall. When using wall-functions, therefore, it is necessary to ensure that the node next to the wall is sufficiently remote for viscous effects there to be unimportant. Then, for example, the so-called "law of the wall" provides a relation between the local mean velocity parallel to the wall u and the wall shear stress τ_s. The following generalised form presented by Launder and Spalding [34] is adopted:

$$u/(\tau_s/\rho)^{1/2} = \frac{1}{\kappa} \ln \left[(Ey_{++})(\tau_s/\rho)^{1/2}/k_p^{1/2} c_\mu^{1/4}\right] \tag{2.50}$$

where κ and E are constants taking values 0.41 and 9.0 respectively and y_{++} is defined as $(y\rho k_p^{1/2} c_\mu^{1/4}/\mu)$, y being the distance from the wall and k_p the value of k at the node of a cell P touching the wall. As u can be evaluated from the stream function distribution, this equation allows τ_s to be computed. The corresponding formula for turbulent viscosity is:

$$\mu_t/\mu = \kappa y_{++} \; . \tag{2.51}$$

When diffusion and convection of kinetic energy are negligible, the conservation equation for k implies that $k_p c_\mu^{1/2}$ equals τ_s/ρ. Under these conditions, (2.50) reduces to the familiar form of the "law of the wall" while Eq. (2.51) assumes the form:

$$\mu_t/\mu = \kappa y\sqrt{\rho\tau_s}/\mu \; . \tag{2.52}$$

The reason for preferring Eq. (2.51) in the present calculations is that in regions of vanishingly small τ_s (for example near a stagnation point) it gives plausible values of μ_t whilst those indicated by (2.52) are unrealistically low.

At the wall itself velocity fluctuations vanish and k becomes zero. This can be used as a boundary condition for k if the finite-difference grid is sufficiently fine to give adequate coverage in the laminar sub-layer. However, the present turbulence model is not applicable in this region and, as for the velocity field, the first node removed from the wall must be situated in the wholly turbulent region. For such a near-wall cell the generation and dissipation of k are far more significant than its diffusion through the bounding wall; accordingly zero diffusive transport to the wall is adopted as the boundary condition in the energy-balance calculation for this cell. Moreover, to improve accuracy the k source-terms for this cell are evaluated as the mean over the cell. Precisely parallel considerations apply to the evaluation of g near a wall. For the energy dissipation rate, equations (2.51) and (2.5) were used to determine the variation of ϵ with distance from the wall.

3. THE NUMERICAL CALCULATION OF CONFINED TURBULENT DIFFUSION FLAMES

3.1 *Introduction*

We now turn to the prediction of the flow and thermodynamic properties of the four flames of Table 3.1, with the aid of the mathematical models discussed in the preceding section. The predicted distribution of velocities, turbulence energy, temperature, etc., will be discussed qualitatively only since no suitable experimental data are available; the predicted f will be compared with measurements for flame A (Table 3.1).

The writer [6] has used a method for mapping the probability of chemical reaction, **P**, from measurements of the instantaneous positive-ion current in the furnace; comparison of these measurements with predictions is provided in Sec.3.6. In order to predict **P**, some assumptions additional to those for the chemical-reaction model are necessary; these will be discussed in Sec.3.4.

3.2 *The Problem Specification*

The flow considered is that of a turbulent diffusion flame confined in the cylindrical furnace shown in Fig.4. There is no swirl in either the fuel or the

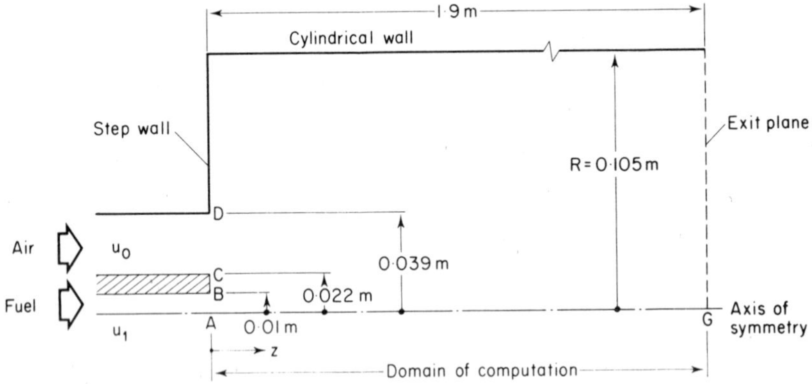

Fig.4 **The region of numerical computation.**

air stream. For simplicity, heat transfer to or from the system is ignored as discussed in Sec.2.2.

Equations set of the type (2.3) are solved simultaneously for the six dependent variables ψ, ω/r, f, k, ϵ and g together with the auxiliary relations

Table 3.1 The test conditions for the predictions.

Flame identifying code	I	II	III	IV	A
Experimental Investigator	Elghobashi [6]				El-Mahallawy *et al.* [36]
Town gas vol. analysis:					
H	0.468	0.446	0.468	0.486	0.55
CH_4	0.29	0.323	0.29	0.309	0.27
CO_2	0.102	0.101	0.102	0.111	0.08
N_2	0.073	0.075	0.073	0.049	0.04
CO	0.027	0.023	0.027	0.024	0.04
C_2H_6	0.04	0.032	0.04	0.021	0.02
Stoichiometric A/F ratio	9.5	9.6	9.5	9.7	10.6
Inlet A/F ratio	22.9	15.4	20.1	16.9	15.75
Lower value of heat of reaction H (J/kg of reactants)	5.75×10^7	5.77×10^7	5.75×10^7	5.88×10^7	5.75×10^7
Inlet fuel vel. u_1 (m/s)	6.01	17.18	22.0	23.0	21.57
Inlet air vel. u_o (m/s)	17.37	12.64	21.0	17.0	13.46
Inlet vel. ratio (u_1/u_o)	.9	1.4	1.05	1.32	1.6
Inlet fuel density ρ_1 (kg/m^3)	0.566	0.572	0.566	0.542	0.474
Inlet air density ρ_o (kg/m^3)	1.165	1.165	1.165	1.165	1.165
Re based on inlet conditions and furnace diameter	20500	15200	25000	21200	16200
Fuel nozzle i.d. (m)	.020	.020	.020	.020	.020
Fuel nozzle o.d. (m)	.045	.045	.045	.045	.045
Air nozzle i.d. (m)	.078	.078	.078	.078	.078

for μ_{eff} and ρ, (Eqs (2.5) and (2.47) respectively). The boundary conditions for each of the six dependent variables are discussed below.

Fuel and air inlets AB and CD. Uniform axial-velocity profiles, u_1 and u_o, are prescribed for the fuel and air streams respectively (see Table 3.1). The radial velocity v_o and v_1 are both zero. The stream function distribution is determined from the definition of ψ (2.1). The values of k and ϵ are prescribed as follows:

$$k_o = .005u_o^2\,, \qquad k_1 = .005u_1^2\,,$$

$$\epsilon_o = k_o^{1.5}/\ell_o \text{ and } \epsilon_1 = k_1^{1.5}/\ell_1\,,$$

where ℓ_1 and ℓ_o are the length scales of turbulence in the two streams, taken as .011 times the radius of AB and the annular gap CD respectively. From the definitions (2.12) and (2.13), the f values are unity and zero in the fuel and air streams respectively. Because of the uniform distribution of f in each of the streams, the g values are zero on the inlet plane.

The separation lip BC. On the lip BC, ψ is a constant, ψ_B; also the normal gradient $\partial\psi/\partial z$ is zero, as is that of f and g. Boundary conditions for k, ϵ and ω/r are evaluated with the aid of the wall functions discussed in Sec.2.3.

The step wall DE and the cylindrical wall EF. The treatment of the boundary conditions for all variables along both walls is the same as that described above for BC; here however $\psi_{DE} = \psi_{EF} = \psi_D$.

The exit plane FG. The assumption of zero axial-gradient for all variables at the exit plane is adopted here.

The axis of symmetry AG. The stream function is set to zero while the radial gradients of all other quantities are made zero.

3.3 *Values for the Coefficients in the Turbulence Model*

The values of the constants c_1, c_2, c_{g1}, c_{g2}, c_μ, σ_k and σ_ϵ appearing in the equations for k, ϵ and g are given in Table 3.2. Those appearing in the hydrodynamic turbulence model (c_1, c_2, c_μ, σ_k and σ_ϵ) correspond with values recommended by Jones and Launder [19]. A re-optimised set has subsequently been reported [34] though it is unlikely that for present purposes there would have been significant differences if that set had been used.

Table 3.2. Coefficients in the turbulence model.

c_1	c_2	c_{g1}	c_{g2}	c_μ	σ_k	σ_ϵ	σ_f	σ_g
1.45	2.	2.8	2.	.09	1.	1.3	.6	.6

The source and sink coefficients in the species fluctuation equation (c_{g1} and c_{g2}) are equivalent to values proposed by Spalding [27] while the values of σ_g and σ_f are only slightly smaller.

3.4 *Calculating the Local Probability of Chemical Reaction,* **P**

(a) The relation between **P** *and* $\tilde{f}$. As mentioned in Sec.3.1, some additional assumptions other than the chemical-reaction model (Sec.2.4) are needed to predict **P**. These are now discussed.

At a given point in the flame, the probability that $\tilde{f}$ lies between two values, $\tilde{f}_a$ and $\tilde{f}_b$ is

$$\int_{\tilde{f}_a}^{\tilde{f}_b} P(\tilde{f})\, d(\tilde{f}) \tag{3.1}$$

This provides the basis for determining **P**. Since reaction is possible only when $\tilde{f}$ is close to f_{st}, we shall suppose at the outset that $\tilde{f}_a$ and $\tilde{f}_b$ should be symmetrically situated on either side of f_{st}, subject to the conditions:

(i) $\tilde{f}_a \geqslant 0$ and $\tilde{f}_b \leqslant 1$,
(ii) when $\tilde{f}_a$ or $\tilde{f}_b$ take on their limiting values, the delta functions

A_o or A_1 (see Fig.3) should be excluded from $P(\tilde{f})$. By introducing a "reaction range" λ indicated on Fig.3, the probability of reaction, **P**, can be written as:

$$\mathbf{P} \equiv \int_{f_{st}-\lambda}^{f_{st}+\lambda} e^{-(\tilde{f}-F)^2/2\sigma^2} \, [U(\tilde{f}-0) - U(\tilde{f}-1)] \, d\tilde{f}/\sigma\sqrt{2\pi} \quad . \tag{3.2}$$

P is represented by the shaded area under the probability density curve of Fig.3. Thus, once λ is known, **P** can be evaluated.

(b) The reaction range, λ. At first an optimum constant value for λ was sought that would give satisfactory agreement between the predicted and measured values of **P**. This attempt was guided by the concept of flammability limits of combustible mixtures. From Eq. (3.2) for a constant λ, **P** is a function of f and g only. A consequence of this is that **P** will have the same value at two separate locations in the furnace if the values of f and g are the same at these points. Measurements indicate, however, that **P** values can be very different at two separate points having the same predicted levels of f and g; it was thus concluded that λ could not be assumed constant.

Having decided that λ should be allowed to vary, the question arose as to what parameters it should be deemed to depend on. For simplicity it was held to depend only on quantities that were already available and only relatively simple functional relationships were considered. Because the experimental values of **P** were deduced from measurements of local ion-concentrations [6], it was felt that consideration of the chemical kinetics of the ionization process would give some useful information.

Let us write the three chemical reactions describing the chemi-ionization mechanism as follows:

chemi-ionization

$$CH + O \xrightarrow{k_i} CHO^+ + e^- , \qquad \Delta H = 0 \text{ kJ/mol} \tag{3.3}$$

charge exchange

$$CHO^+ + H_2O \xrightarrow{k_e} H_3O^+ + CO , \qquad \Delta H = -34 \times 4.187 \text{ kJ/mol} \tag{3.4}$$

ion recombination

$$H_3O^+ + e^- \xrightarrow{k_r} H_2O + H , \qquad \Delta H = -145 \times 4.187 \text{ kJ/mol} \tag{3.5}$$

where k_i, k_e and k_r are the reaction-rate constants for the reactions in question. It can be shown that the rate of production of H_2O, $\dfrac{dH_2O}{dt}$ *, is

* The rate of H_2O production from the ion-recombination reaction is in fact only a small fraction of the total H_2O production rate.

approximately proportional to n_+^2 where n_+ is the local instantaneous molar concentration of the positive ion H_3O^+. A connection between n_+^2 and other quantities was sought.

The net rates of production for H_3O^+ and of recombination for CHO^+ can be written:

$$\frac{d[H_3O^+]}{dt} = k_e[CHO^+][H_2O] - k_r[H_3O^+][e-] \tag{3.6}$$

$$\frac{d[CHO^+]}{dt} = k_i[CH][O] - k_e[CHO^+][H_2O] \tag{3.7}$$

where $[j]$ denotes the instantaneous molar concentration of the species j.

In a steady state, the net rates $\frac{d[H_3O^+]}{dt}$ and $\frac{d[CHO^+]}{dt}$ must equal zero. Thus,

$$[H_3O^+][e^-] = \frac{k_i}{k_r}[CH][O] \; . \tag{3.8}$$

Since the charges are in balance, we have:

$$[H_3O^+] = [e^-] = [n_+] \; .$$

Equation (3.8) therefore becomes:

$$[n_+^2] = \frac{k_i}{k_r}[CH][O] \; . \tag{3.9}$$

Equation (3.9) states that the square of the instantaneous positive-ion concentration is proportional to the product of the instantaneous concentrations of the radicals CH and O. The proportionality constant, k_i/k_r, can be evaluated from the Arrhenius relation with the minimum activation energies assumed to equal the exothermisities of the reactions (3.3) and (3.5) respectively (Green and Sugden [36]). Thus,

$$\frac{k_i}{k_r} = \frac{A_i}{A_r} e^{-E/\mathcal{R}T} , \tag{3.10}$$

where A_i and A_r are the pre-exponential coefficients, and T is the time-mean temperature of positive-ions, which is nearly equal to the local gas temperature (Calcote and King [37]). The activation energy E equals 145×4.187 kJ/mol (Eqs 3.3 and 3.5) and $\mathcal{R}$ is the universal gas constant. The ratio (A_i/A_r) can be replaced by a constant, A, since both A_i and A_r are proportional to $T^{1/2}$. Thus, expression (3.9) can be written:

$$[n_+]^2 = Ae^{-E/\mathcal{R}T}[CH][O] \; . \tag{3.11}$$

Now $[n_+]^2$ can be taken proportional to the reaction rate; thus we may expect that λ can be related to the quantities on the right-hand side of Eq. (3.11). Since, in the present prediction procedure the concentration of the radicals such as CH and O is not calculated, it is necessary to relate their concentrations to other available quantities. To do this, additional assumptions must be introduced.

For convenience, let us re-write Eq. (3.11) as:

$$\left(\frac{d}{dt}[H_2O]\right)_i^* \approx Ae^{-E/\mathcal{R}T}\,[CH]\,[O] \;, \tag{3.12}$$

where the approximation sign comes from the relation between $[n_+]^2$ and $\left(\frac{d}{dt}[H_2O]\right)_i$. Let the symbols $\tilde{\alpha}$, $\tilde{\beta}$ and $\tilde{\gamma}$ stand for the instantaneous concentrations of $[O]$, $[CH]$ and $([H_2O])_i$ respectively. We can then write Eq. (3.12) as:

$$\frac{d}{dt}\tilde{\gamma} \approx \tilde{\alpha}\tilde{\beta}Ae^{-E/\mathcal{R}T} \;. \tag{3.13}$$

If we split the instantaneous quantities $\tilde{\alpha}$, $\tilde{\beta}$ and $\tilde{\gamma}$ in conventional fashion, as:

$$\tilde{\alpha} = \alpha + \alpha' \;; \qquad \tilde{\beta} = \beta + \beta' \;; \qquad \tilde{\gamma} = \gamma + \gamma' \;,$$

substitute them in Eq. (3.13) and time-average we get:

$$\frac{d}{dt}\gamma \approx Ae^{-E/\mathcal{R}T}\,(\alpha\beta + \langle u'\beta'\rangle) \;. \tag{3.14}$$

Equation (3.14) highlights the important role the concentration fluctuations play in controlling the chemical reaction rates. In particular, the reaction rate can either be reduced or enhanced by the correlation term $\langle \alpha'\beta'\rangle$ depending on whether it is negative or positive.

As an illustration, imagine two reacting materials $\tilde{\alpha}$ and $\tilde{\beta}$ which pass a point of observation at different times (i.e. they are never in contact with each other). Time-mean values α and β are measured. The product $\alpha\beta$ is equal to $-\langle \alpha'\beta'\rangle$ which in this case is zero; thus no chemical reaction is predicted. On the other hand, if the two materials are initially premixed and discharge from a point source, during their subsequent turbulent mixing with the surrounding atmosphere their concentration fluctuations should be positively correlated. We should then expect, on the basis of (3.14) that their reaction rate would be considerably higher than if there were no fluctuations, or if the flow was non-turbulent.

In the above discussion, the important role of molecular diffusivity in dissipating the correlation $\langle \alpha'\beta'\rangle$ has been neglected. This decay of concentration fluctuations is analogous to the decay of turbulence kinetic energy; the

* Subscript i denotes here the rate of production of H_2O from ion-recombination.

former is due to molecular diffusion fine scale motions while the latter arises from molecular viscosity. However, the time-rates of decay of both quantities are controlled by large-scale dynamics which involve neither molecular diffusion nor viscosity (Batchelor [38], Csanady [26], Tennekes and Lumley [39]). It is basically for this reason that the dissipation term in the g equation (Sec.2.2) has been modelled as proportional to $g\epsilon/k$ (Table 2.2) where ϵ/k is the time-scale of the large turbulence eddies.

It may thus be expected that the time-mean rate of reaction $d\gamma/dt$ in Eq. (3.13) would be proportional to the decay rate of $\alpha'\beta'$. To calculate this rate rigorously, five partial differential equations for the dependent variables $<\alpha'\beta'>$, $<\alpha'^2>$, $<\beta'^2>$, α and β would have to be solved simultaneously with the other transport equations. Such an approach has been suggested by Hilst *et al.* [40]. In the present treatment, the mathematical models for the chemical reaction and concentration fluctuations allow us to calculate α, β, $<\alpha'^2>$ and $<\beta'^2>$, but no information about $<\alpha'\beta'>$ can be obtained. For this reason we shall postulate here that the rate of decay of $<\alpha'\beta'>$ is proportional to that of g; this permits the rate of reaction, $d\gamma/dt$, to be taken as proportional to dg/dt.

As indicated earlier, the "reaction range" λ will be related to the quantities on the right-hand side of (3.11). To be concrete we shall now assume λ to be proportional to both $[Ae^{-E/\mathcal{R}T}]$ and $[dg/dt]$. Since λ is dimensionless, the term $[dg/dt]$ needs to be modelled in dimensionless form. One way of doing this is to write $[dg/dt]$ as $[g/f^2][t_K/t]$, where t equals ϵ/k and t_K is the Kolmogorov time-scale, $(\nu/\epsilon)^{1/2}$. The Kolmogorov time-scale is proportional to the lifetime of the smallest scales occurring in turbulent motion; it decreases with increasing dissipation rate of turbulent energy. Thus a fairly general expression for λ may be written as:

$$\lambda = Ae^{-E/\mathcal{R}T}\,[g/f^2]^{B_1}\,[t_K/t]^{B_2}\ , \tag{3.15}$$

where B_1 and B_2 are constants of proportionality. Their values, together with that of A have been chosen by correlating the measured and predicted values of **P** at many locations in the furnace (150 points from 4 sets of results), using the least square method. The values obtained* were:

$$A = 1.2\ , \qquad B_1 = 1\ , \qquad B_2 = 1\ .$$

It may be noted that Eq. (3.15) can be written in terms of length-scales instead of the time-scales as:

$$\lambda = Ae^{-E/\mathcal{R}T}\,[g/f^2]^{B_1}\,[\ell/\eta]^{2B_2/3}\ , \tag{3.16}$$

* These values differ slightly from those published by Elghobashi and Pun [41] (where they were given as 1.1, 1.1, 1. respectively). The reason is that at the time of that publication, the experimental results of flames III and IV were not included in the optimisation.

where ℓ is the turbulent length-scale proportional to the size of large eddies (i.e. (2.8), and η is the Kolmogorov length scale, $\nu^{3/4}/\epsilon^{1/4}$.

3.5 *Numerical Accuracy, Convergence Criteria and Computer Time*

Computations were performed with 15X15 and 20X30 non-uniform grids (with the larger number of nodes in the radial direction). Computer storage required for the finer mesh was about 60K, which was the maximum available to the writer on the ULCC CDC-6600.

Figure 5 shows the effect of the grid size on the distribution along the furnace axis of f and T for flame A. (Also shown on Fig.5 are the experimental values of f (El-Mahallawy *et al.* [35] for the same flame.) On the whole, the results from the coarse and the fine grids do not differ significantly, except perhaps at locations close to the burner where the gradients in both the axial and radial directions are very steep. On this basis the 15X15 grid was considered accurate enough and all computations reported later were carried out with this more economical grid.

Convergence of the calculation was measured by reference to two criteria. The first was the monotonic decrease of the quantity $\Delta\phi$ defined as:

$$\Delta\phi \equiv \frac{\phi^n - \phi^{n-1}}{\phi_{ref}} , \qquad (3.17)$$

where ϕ^n and ϕ^{n-1} denote the values at a node of variable ϕ at two successive iterations and ϕ_{ref} is a suitable reference value of ϕ. The second criterion was based on the use of a residual source R_s defined as:

$$R_s \equiv \left|\phi_P - \sum_{i=N,S,E,W} c_i\phi_i - d_\phi\right| , \qquad (3.18)$$

where ϕ_P is the value of ϕ at the grid node P; N, S, E, W are the neighbouring nodes; the c_i's are coefficients representing the influence of convective and diffusive fluxes, and d_ϕ is the source term. When the maximum value of R_s for any cell had fallen below some suitably small value the solution could be regarded as converged.

The computer time required depended on the number of grid points and the number of iterations necessary for convergence. For a 15X15 grid, about 400 iterations were needed which required approximately 540 s central processor time of a CDC-6600 computer. For a 20X30 grid 1140 s were needed for 446 iterations.

3.6 *Presentation and Discussion of Results*

We now consider the results of some of the predictions for flames I, II, III, IV and A of Table 3.1. The different inlet conditions of these flames are chosen so as to allow the effects of Reynolds number, air/fuel ratio and velocity ratio to be studied. Where no suitable measurements are available, the computations are examined qualitatively.

The results for flames I-IV are discussed in Figs 5–8. The observations

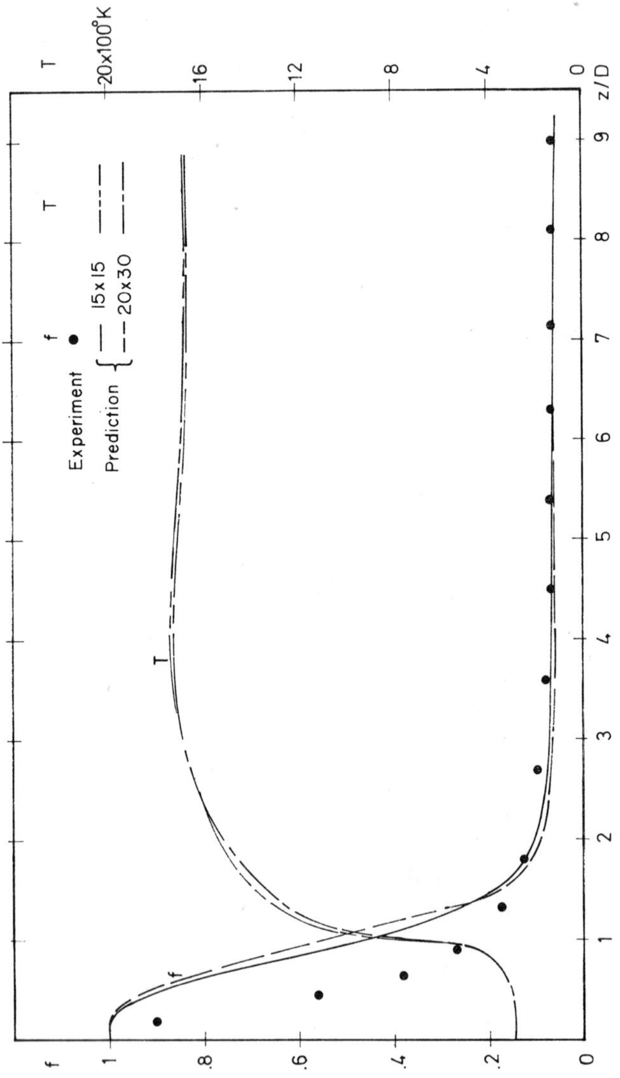

Fig.5 Variation of mixture fraction and temperature along furnace axis — flame A.

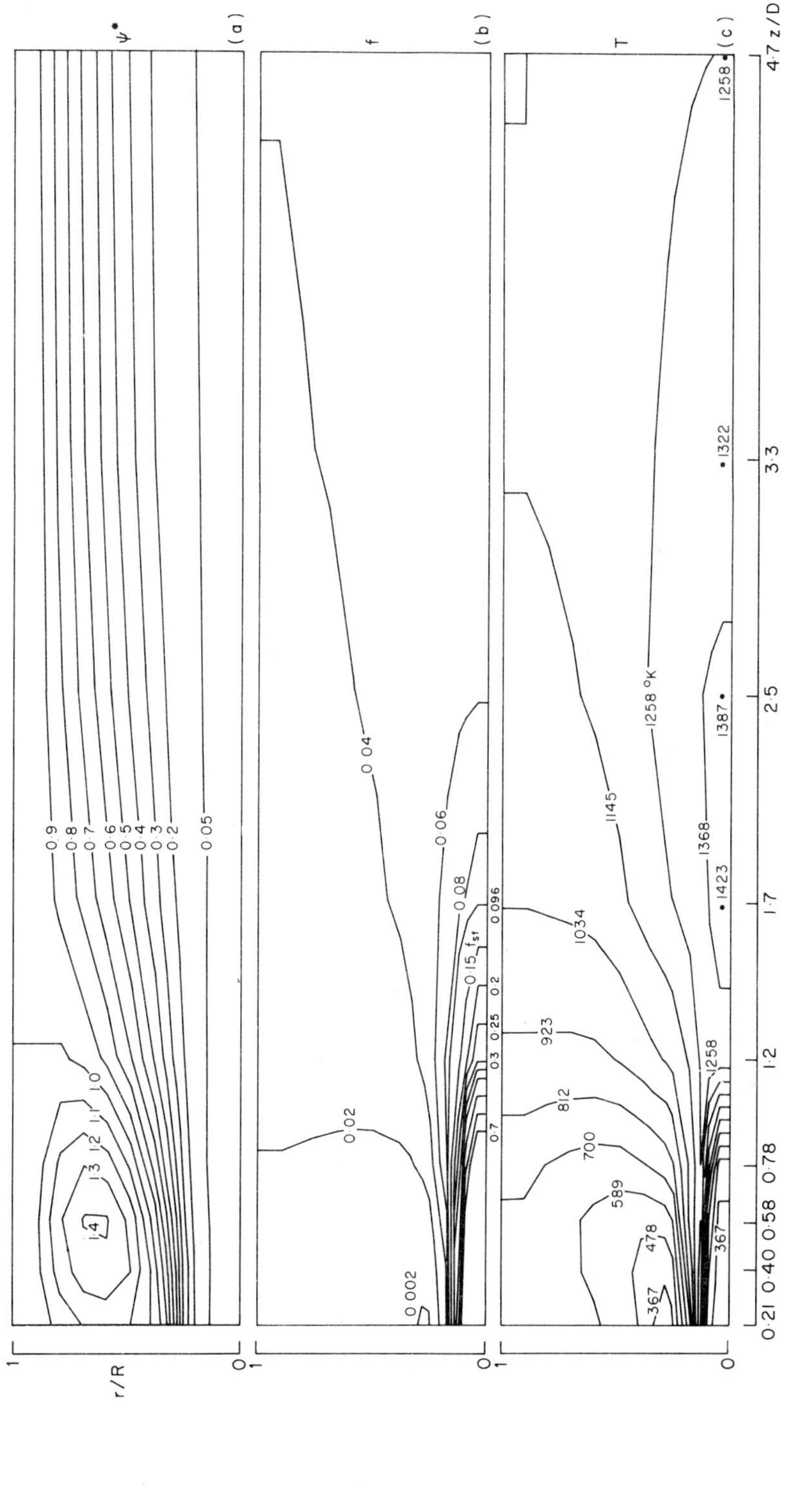

Fig.6 Predicted characteristics of flame I. *a.* Normalised stream function. *b.* Mean mixture fraction. *c.* Mean temperature.

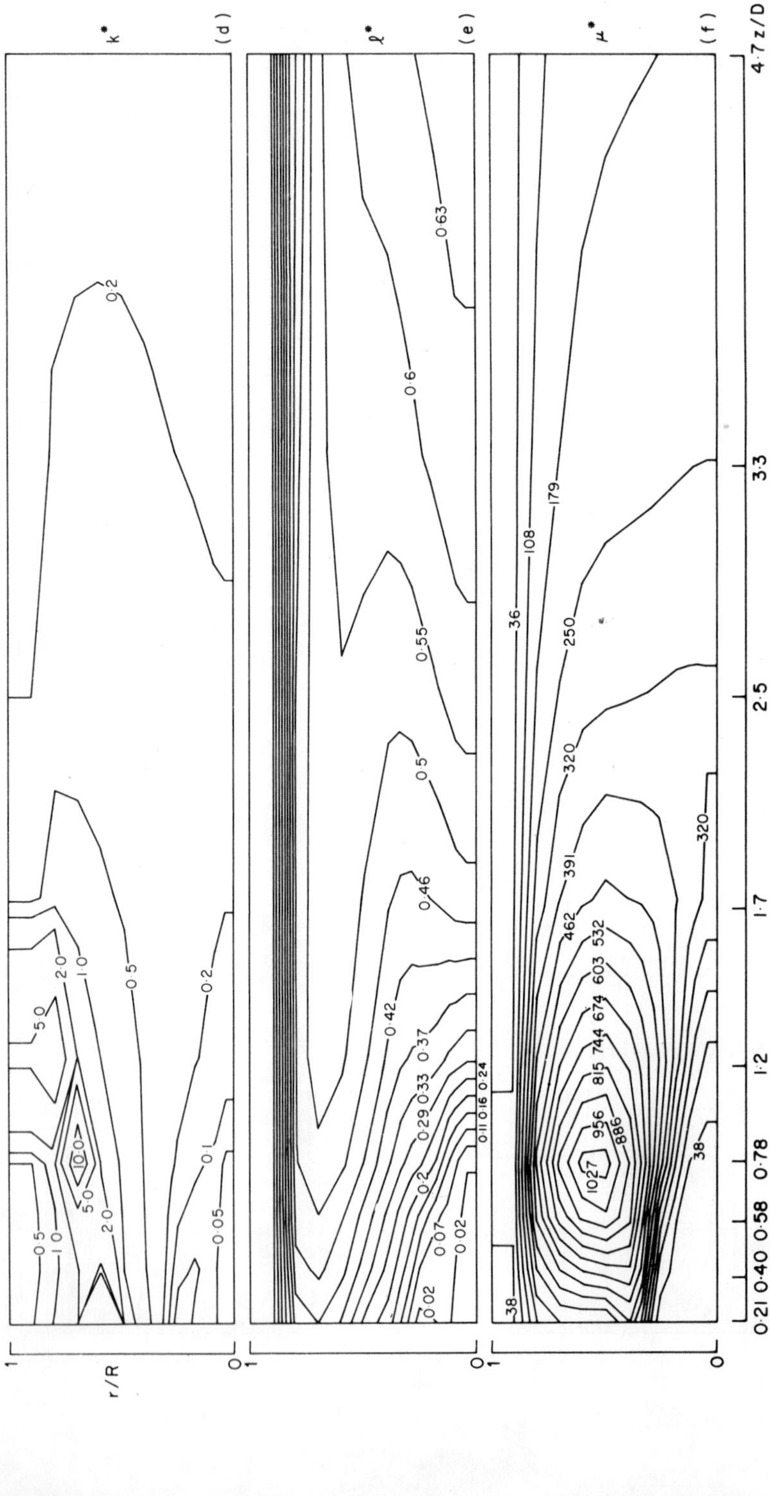

Fig. 6 (cont.) *d.* Turbulence kinetic energy. *e.* Turbulence length scale. *f.* Effective viscosity.

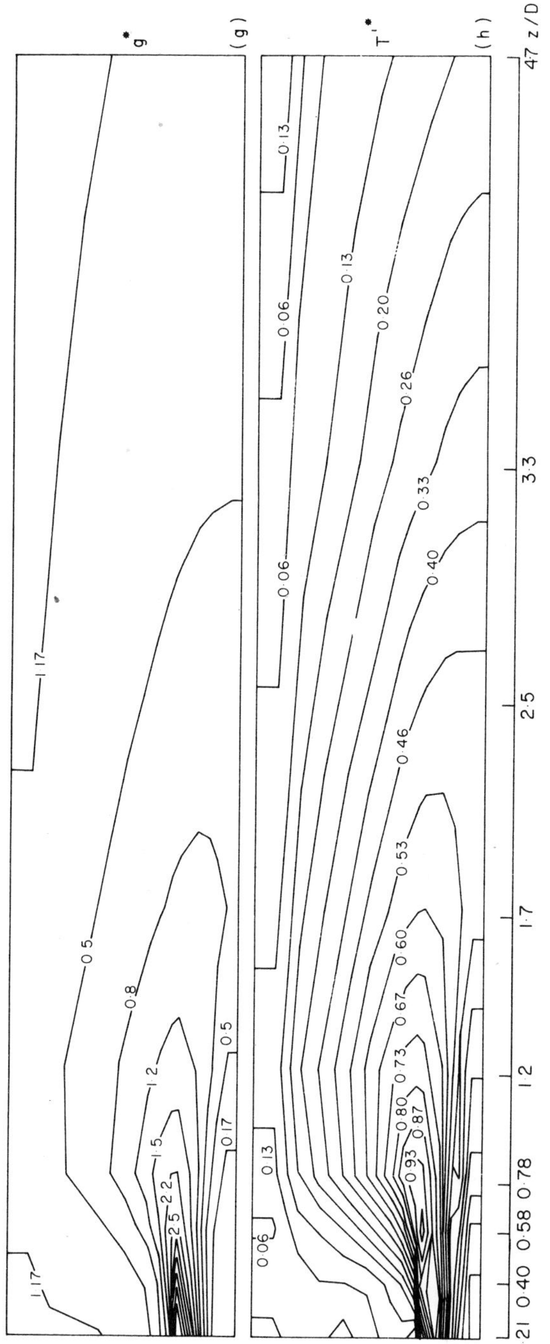

Fig.6 (cont.) *g.* Intensity of mixture fraction fluctuations. *h.* Intensity of temperature fluctuations.

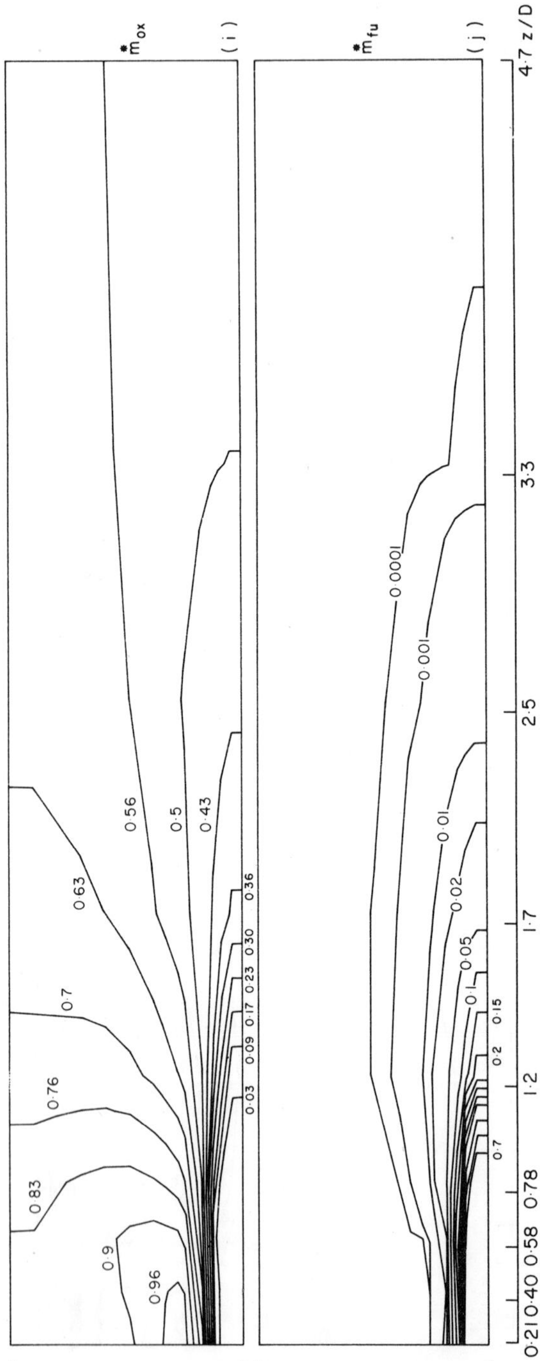

Fig.6 (cont.) *i.* Mean mass fraction of oxygen. *j.* Mean mass fraction of fuel.

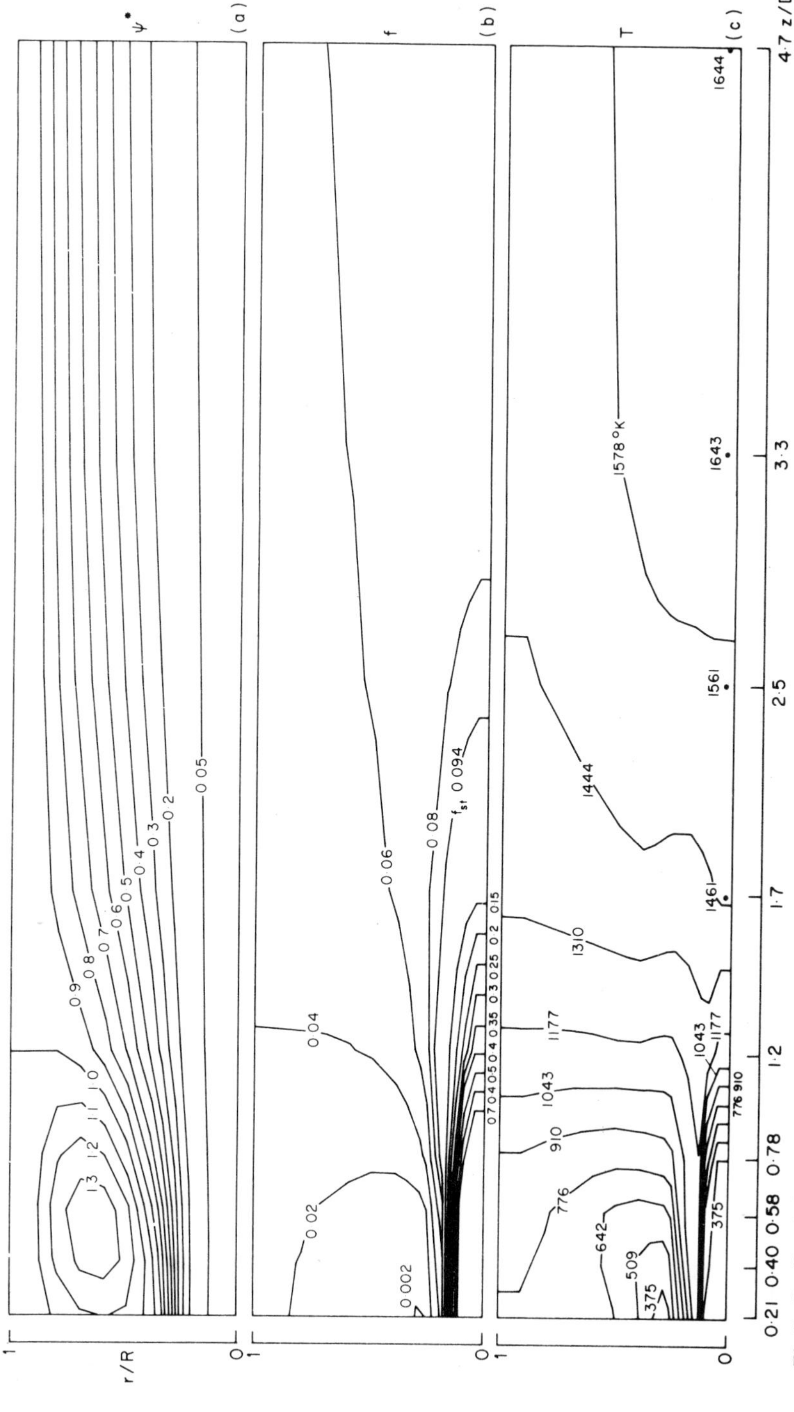

Fig. 7 Predicted characteristics of flame II. *a*. Normalised stream function. *b*. Mean mixture fraction. *c*. Mean temperature.

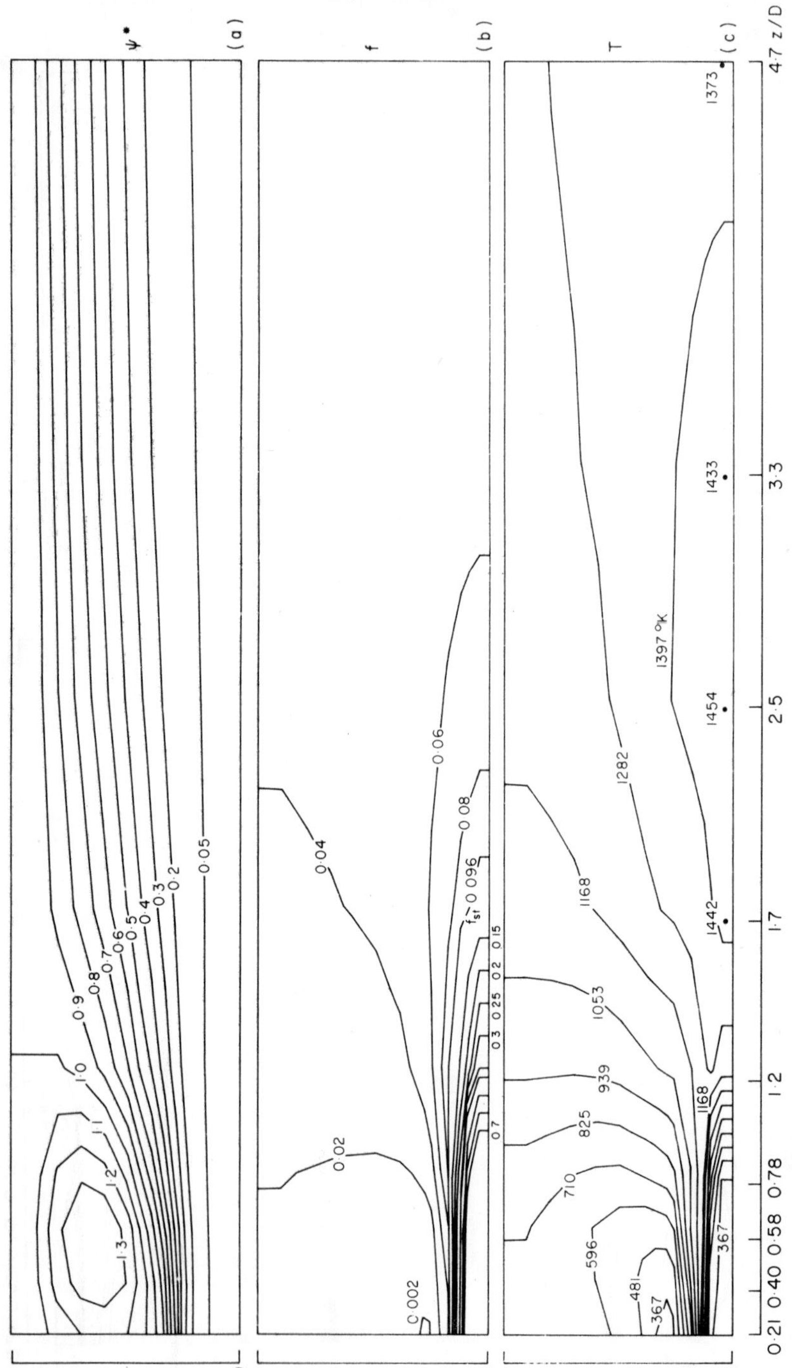

Fig.8 Predicted characteristics of flame III. *a.* Normalised stream function. *b.* Mean mixture fraction. *c.* Mean temperature.

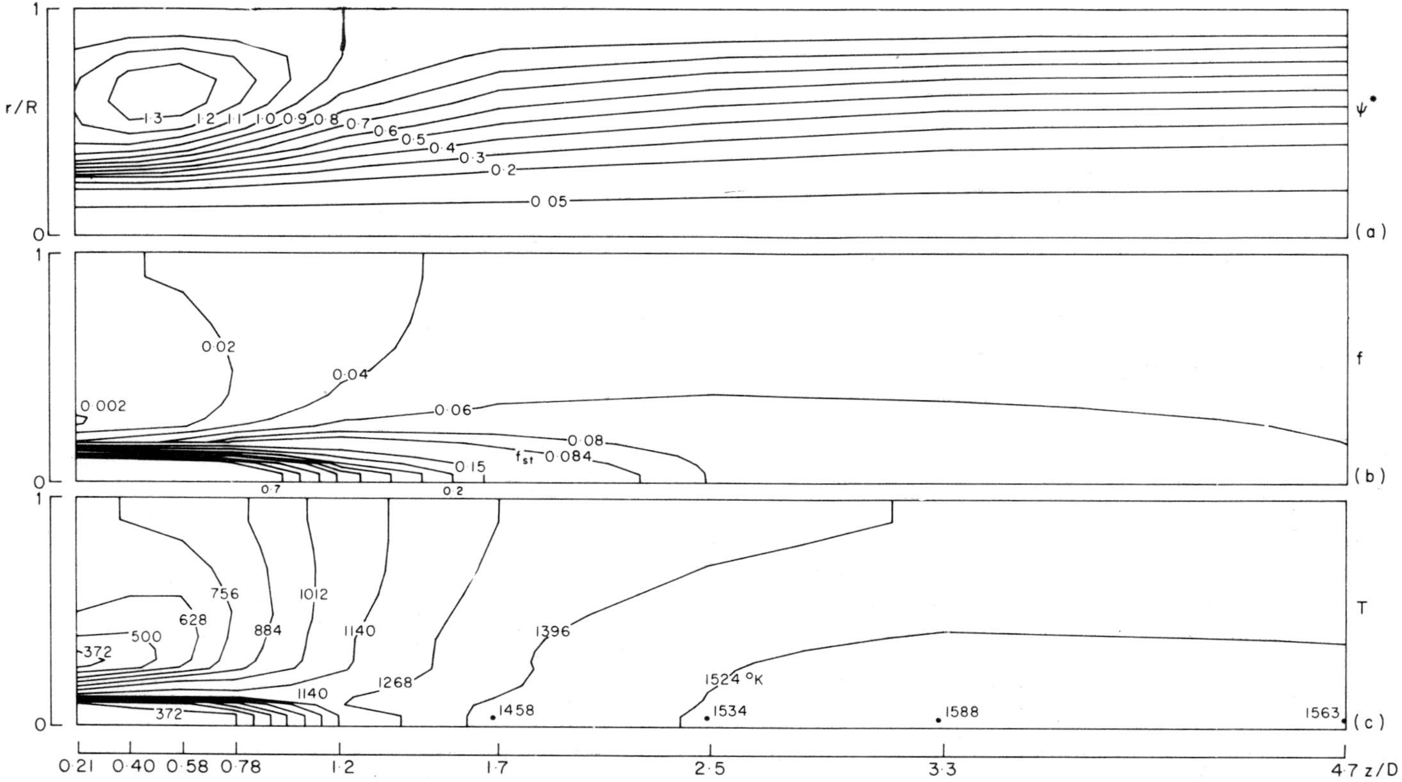

Fig.9 Predicted characteristics of flame IV. *a.* Normalised stream function. *b.* Mean mixture fraction. *c.* Mean temperature.

made apply to all four flames unless specifically stated otherwise. All quantities plotted, except the temperature are dimensionless. The reference value of the stream function ψ_{ref} is the maximum value at the inlet plane of the combustion chamber (i.e. it is proportional to the sum of the mass flow rates of fuel, oxidant and inert species).

(a) The stream function. From Figs 6a, 7a, 8a and 9a it is seen that the length of the recirculation zone is nearly the same, about 1.2D, for the four different conditions. Moreover, the maximum mass-flow rate of the recirculating products of combustion is about 40% of the total inlet mass flow rate. For the range of inlet conditions studied (Re varied from 15200 to 25000) the predictions suggest that, as in isothermal flows, the length of the recirculation zone in a turbulent, sudden-expansion, reacting flow is mainly a function of the expansion ratio.

(b) Turbulence energy and length scale of turbulence. It may be seen from Fig.6d that there are two locations of very high turbulence intensity ($k^* \equiv k^{1/2}/u \geqslant 5$). The first is located at an axial distance of about 0.8D from the burner and at a radial position of about 0.7R; the second is closer to the wall and at an axial distance of 1.2D. As expected, these two zones are where $\partial u/\partial r$ is greatest (thus generation rates of k are large) and the mean velocity is small; they occur at the centre and close to the reattachment of the recirculating eddy.

The normalised turbulence length-scale ℓ^* ($\equiv \ell/R$) shown in Fig.6e exhibits a universal distribution near the wall; this is a consequence of the wall-function treatment (Sec.2.3), which at the nearest node to the wall makes the length scale proportional to the distance from the wall.

A region of high turbulence dissipation, characterised by small values of the length scale ($\ell^* = .027$ to $.07$), exists between the fuel and air streams ($r/R = 0.1$ to 0.25) close to the burner (for $z/D \leqslant 1$). As will be seen later this region is particularly significant in promoting chemical reaction. The variation of the length scale along the axis shows a steep increase from $\ell^* = 0.02$ at $z = 0.78D$ to $\ell^* = 0.29$ at $z = 1.2D$; these two locations mark the axial positions of the center and the end of the recirculating eddy. Downstream from the zone of recirculation, the length scale continues to increase, but at a much slower rate, to the end of the furnace where its value exceeds 0.8R.

(c) The effective viscosity. The effective normalised viscosity (μ_{eff}/μ) shown in Fig.6f reaches its highest levels ($\mu \approx 10^3$) in a zone located at an axial distance of about 1.5R from the burner and at $r = 0.5R$. This is a region of high generation of k due to the steep velocity gradients present there. The large values of μ_{eff} (and hence of other exchange coefficients) in this region help to emphasize the important role of the recirculation of hot combustion products in heating the incoming air – thereby helping to stabilize the flame.

The computed exchange coefficients for the mixture fraction f in the recirculation zone vary between .03 and .04 m^2/sec; these values agree with the measurements of Lenze [42] for a confined, turbulent diffusion flame of town gas.

(d) Time-mean mixture fraction. In Fig.5, the predicted distribution of f along the axis of the furnace for flame A is compared with the measurements of El-Mahallawy *et al.* [36] for the same conditions. For $z/D \geqslant 1.5$ good agreement is observed. Close to the inlet where the gradients in both the axial and radial directions are steep, the measurements exhibit a faster axial decay-rate than the predictions. It is possible that measurements in this region were subject to larger errors because of the critical need to locate the sampling probe more accurately there. The measured and predicted radial profiles are compared in Fig.10 at three axial locations (z/D equal to 0.21, 1.7 and 2.5). Again agreement between prediction and measurement is good at the two downstream locations. At the station nearest the burner, agreement is less good, particularly close to the axis.

No measurements of f are available for the other flames studied. However, the predictions of f in Figs 6b, 7b, 8b and 9b show that the rate of decay of the mixture fraction with distance along the axis depends mainly on the inlet/fuel ratio. Flame II, the fuel-richest of the four flames, exhibits the lowest rate of decay of f, while flame I, the leanest, shows the highest rate. The rates of decay for flames III and IV fall between these two extremes.

(e) Intensity of fluctuations of the mixture fraction. It is appropriate here to consider the probability density function $P(\tilde{f})$ in parallel with the distributions of the intensity of fluctuations $g^{1/2}/f$ shown in Fig.6g. The intensity $g^{1/2}/f$ reaches its maximum value of about 3* in a region very close to the burner ($z \leqslant 0.4D$) at a radial location of about $0.3R$. In this region, because the level of f is less than 1% $P(\tilde{f})$ will be characterized by large values of the delta function A_o as shown in Fig.11a, that is, the instantaneous mixture fraction $\tilde{f}$ fluctuates between values corresponding to air and products on one side and fresh air on the other, with a large proportion of time spent at the latter.

Close to the axis, where $g^{1/2}/f$ is near unity and f approaches f_{st}, $\tilde{f}$ experiences fluctuations between fuel and products at one end and air and products at the other; the delta function A_o thus diminishes. At the axis itself f takes on large values (≈ 0.5) corresponding to the fuel-rich zone, $g^{1/2}/f$ decreases and $P(\tilde{f})$ approaches a normal Gaussian distribution in which $A_o = A_1 = 0$. An example of such conditions is shown in Fig.11b.

Far downstream from the burner, beyond an axial distance of $4.7D$ up to the end of the furnace, f becomes less than f_{st} and the intensity $g^{1/2}/f$ attains an approximately uniform value of about .16. In that region $P(\tilde{f})$ is expected to have a peaked Gaussian distribution.

(f) Time-mean and fluctuating values of temperature. Figures 6c, 7c, 8c and 9c show that the field of time-mean temperature can be divided into three zones: the first extends from the burner up to $z/D \approx 1.0$; in this zone steep gradients exist in the radial direction and the temperature of the recirculating combustion products (in the region of $r = .5R$ to R) is higher

* This seemingly high value still lies well below the physical upper-limit imposed on g, namely $g \leqslant f(1-f)$.

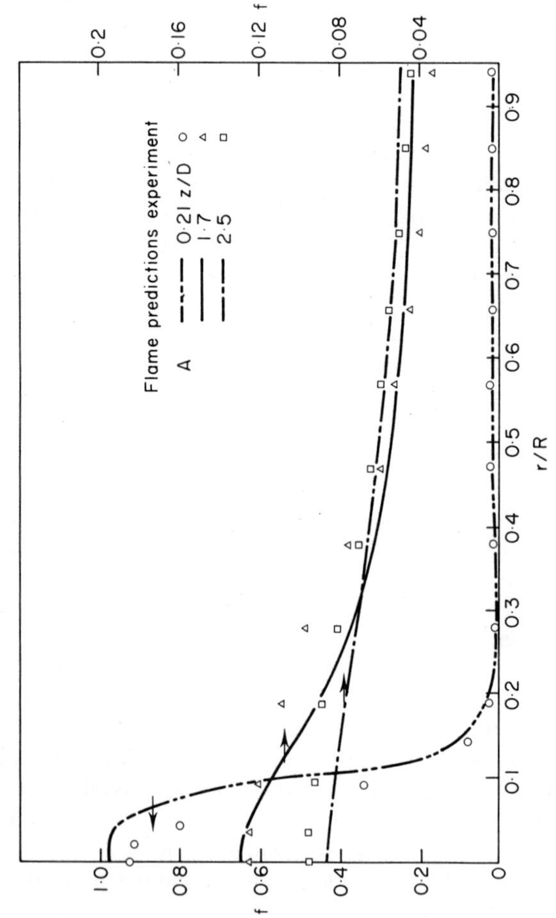

Fig.10 Comparison of predicted and measured profiles of mixture fraction – flame A.

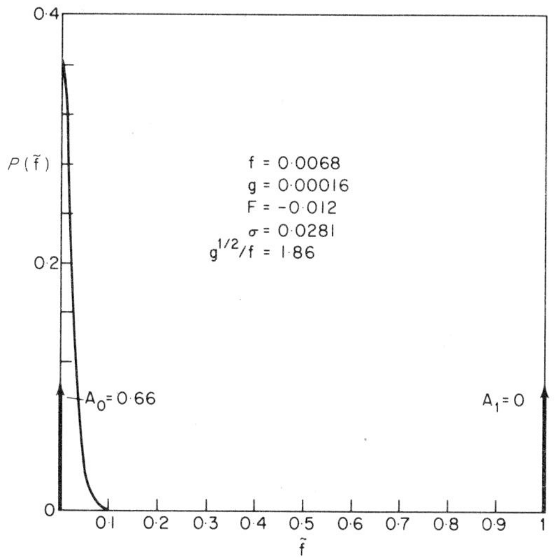

Fig. 11 Distribution of probability density function for two conditions. *a.* Region of high intensity fluctuations near burner discharge.

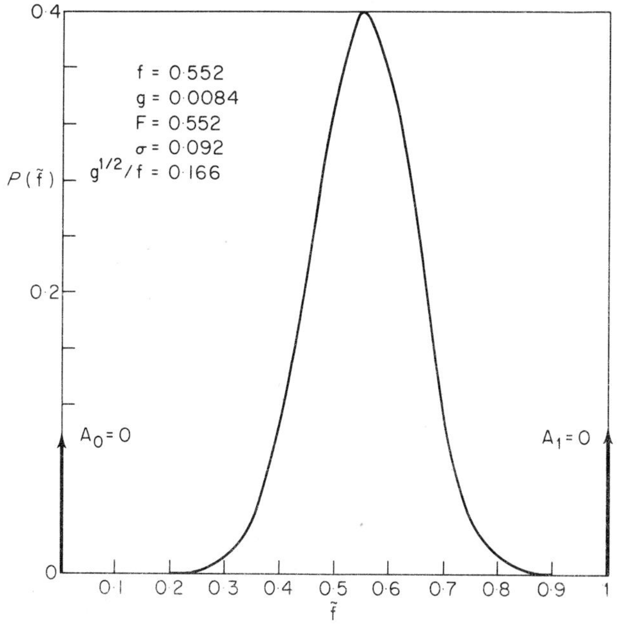

Fig. 11 (cont.) *b.* Conditions near axis, $z/D \approx 1.0$.

than that of the unreacted fuel near the axis. This zone is also characterized by large fluctuations in temperature. From Fig.6h it may be seen that T'^* ($\equiv \langle T'^2 \rangle^{1/2}/T$) is approximately unity there. It can be seen that the maximum fluctuations occur at a larger radial distance $(0.3R)$ from the axis than that of the maximum time-mean temperature $(0.15R)$ where the turbulence length-scale is much smaller. The second zone extends over the axial distance from about $1.0D - 1.7D$; in that region the time-mean temperature along the axis becomes higher than that of the combustion products, but the maximum value of T occurs away from the axis at a radial location of $0.1R$. The temperature fluctuations at the axis exhibit steep gradients because relatively large-scale air vortices (pre-heated by the combustion products) are introduced by the turbulent mixing into the smaller-scale region of the fuel jet. In the third zone which occupies the rest of the furnace the maximum value of T occurs along the axis where the fluctuations in both temperature and concentration decay gradually.

As expected, the time-mean temperature at the exhaust end of the furnace is inversely proportional to the inlet air/fuel ratio, being a maximum for flame I (22.9) and minimum for flame II (15.4).

It is remarked that although the heat transfer to the furnace wall is neglected in the computations the predicted maximum values of the time-mean temperature are, on the whole, 20–30% lower than those measured by Günther and Lenze [44] for a similar flame. This discrepancy may perhaps be due to one (or more) of the following points:

1. Günther and Lenze used a 10 mm diameter fuel-jet in a relatively large furnace (450 mm diameter); thus the flow was essentially a free jet. In contrast strong recirculation exists in cases predicted here which will lead to relatively larger turbulence length scales. A smaller length-scale results in smaller amplitude fluctuations of temperature and consequently in higher time-mean values.
2. The predicted values of the fluctuations of the mixture fraction may be higher than their actual values (the constants c_{g1} and c_{g2} were chosen by reference to non-reacting flows). This would result in lower values for the time-mean temperature.
3. In calculating the time-mean temperature from Eq. (2.46), the instantaneous specific heat was assumed to depend on the local time-mean temperature instead of the instantaneous one. This would lead to an under-estimate of the time-mean temperature at locations where the values of either of the delta functions of $P(\tilde{f})$ were large.

To provide a further, albeit fairly crude, check on the predictions a simple enthalpy balance for the furnace was carried out under the experimental conditions of the four flames. Table 3.3 below displays the four different values of the exhaust temperature (averaged over the exit plane) obtained (a) from the predictions, (b) from a 1-dimensional enthalpy balance for adiabatic walls, (c) from a similar enthalpy balance for non-adiabatic walls and (d) from the measurements using a thermocouple (Comark-type 1602). The predictions are in satisfactory agreement with those from the enthalpy balance calculations for the adiabatic case. The discrepancy between the thermocouple measurements and the non-adiabatic case is thought to be due mainly to thermo-

Table 3.3 Exhaust temperatures.

Flame	Average exhaust temperature °K			
	Predictions adiabatic wall	Enthalpy balance adiabatic wall	Enthalpy balance non-adiabatic	Thermocouple measurements
I	1190	1234	1102	1073
II	1590	1642	1479	1138
III	1300	1400	1277	1088
IV	1500	1560	1404	1173

couple error (no correction for radiation and conduction losses was made).

(g) Time-mean mass fractions of oxygen and fuel. From Figs 6i and 6j we see the expected overlapping near the axis of fuel and oxygen concentrations; this is supported by the currently available experimental data (for example Vranos *et al.* [44]).

(h) Probability of chemical reaction **P**. The predicted distribution of probability of chemical reaction **P** are compared with the writer's measurements [6] in Figs 12 and 13. The figures show the measured probability of reaction to reach maximum levels of about 0.45 for all four flames at about the same axial-position ($z/D \approx 0.4$).

If we take the reaction zone as being the region with values of **P** higher than 10^{-3} we see that close to the burner the measured radial width of this zone is about $0.2R$. As we move downstream from the burner, the radial thickness increases, the inner envelope falls towards and eventually meets the axis, while the outer envelope pushes outwards, reaching a maximum radial distance before falling back towards the axis and reducing the reaction-zone width. This variation agrees qualitatively with that deduced from the reaction-rate profiles measured by Vranos *et al.* [44] and describes the "flame brush" associated with turbulent diffusion flames. It is interesting to note that the reaction zone begins to spread significantly in the radial direction, as expected, only beyond the end of the recirculating zone of hot products at $z/D = 1.2$.

The radial distribution of **P** in Figs 12a–d shows that the predicted values and locations of the maximum **P** agree well with measurements at sections close to the burner but less well farther downstream. The radial width of the reaction zone (i.e. $\mathbf{P} > 10^{-3}$) is predicted within the range of experimental uncertainty (estimated to be of the order of 10%) except at $z/D = 1.2$. At this station the measurements indicate that **P** is only slightly influenced by the different conditions of the four flames, while the predictions show pronounced effect.

The above discrepancies between measurements and predictions downstream from the burner and away from the axis seem likely to be due to the simplifying assumptions made in the formulation of λ, especially in the treatment of the kinetics of ionization which are believed to have significant effects in those regions.

Figure 13 which shows the variation of **P** along the axis for flames I, II, and IV enables us to study the effect of Reynolds number and inlet air/fuel

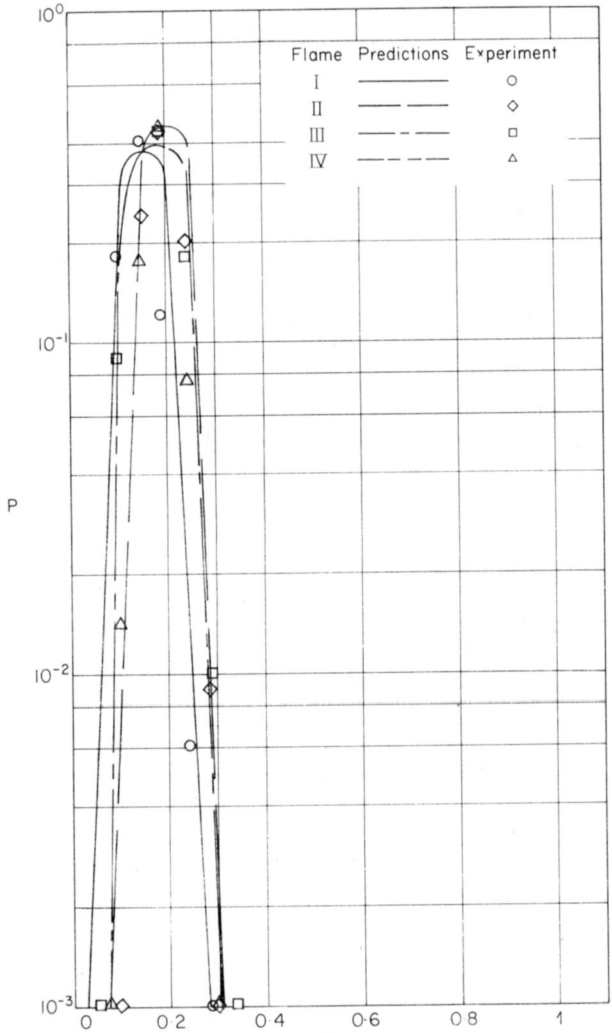

Fig. 12 **Probability of chemical reaction.** ***a.*** **$z/D = 0.40$.**

ratio. Flame II has a Reynolds number of 15200 while that for flame IV is 40% higher (21200). The inlet air/fuel ratios for the two flames (15.4 and 16.9 respectively) differ by only 5%. Figure 13 shows that both these flames reach a value of **P** of .001 at nearly the same axial location, $4.75D$ and $5D$; thus we conclude that the length of the reaction zone is rather insensitive to variations in Reynolds number.

The effect of air/fuel ratios emerges by comparing flames I and IV; the inlet air/fuel ratios for the two are 22.9 and 16.9 respectively. The two flames have about the same Reynolds number. The measured values of **P** indicate

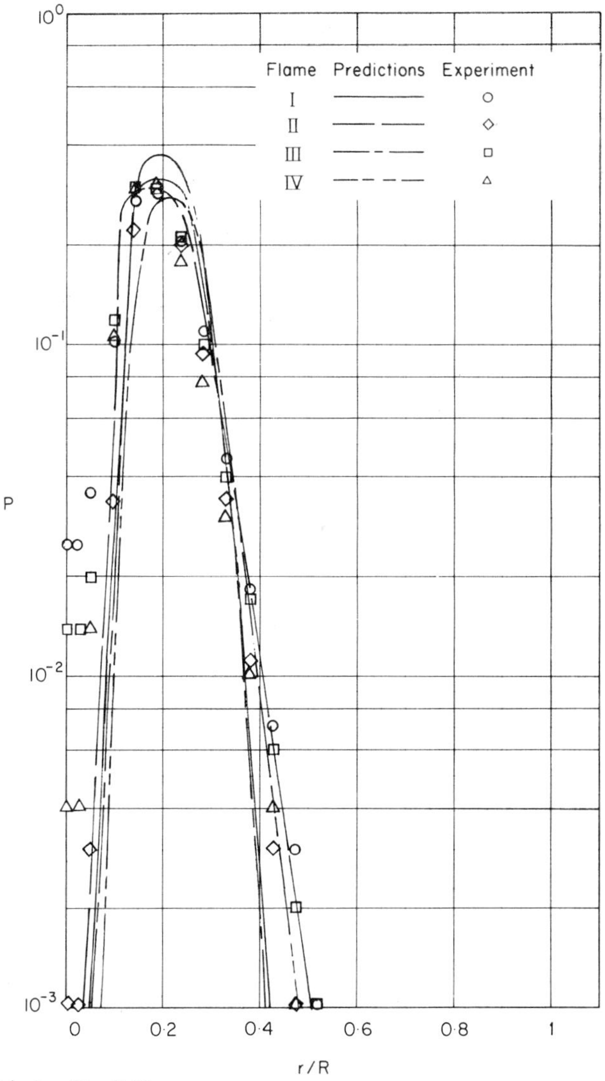

Fig.12 (cont.) *b.* $z/D = 0.78$.

that the reaction zone of flame IV extends $5D$ downstream from the burner, while that of flame I extends only to $3.5D$. Thus increasing the inlet air/fuel ratio shortens the length of the reaction zone.

The agreement between measurements and predictions is on the whole fairly good. The predicted locations of f_{st} on the axis coincide with the predicted (and are very close to the measured) positions of maximum **P** for flames I and II, but in flame IV the measured maximum of **P** occurs nearer the burner.

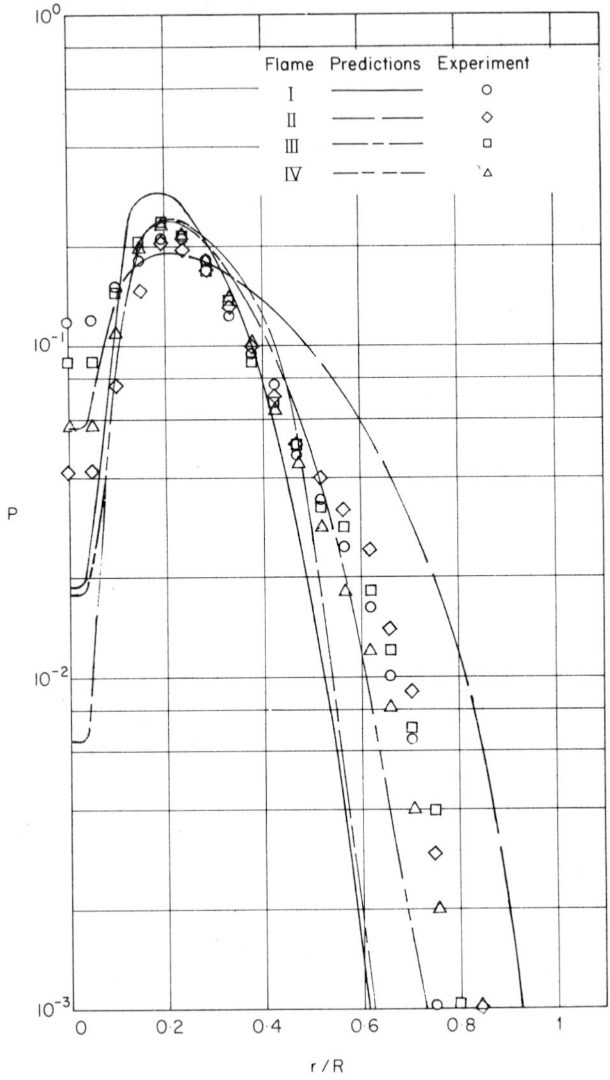

Fig. 12 (cont.) c. $z/D = 1.2$.

4. CONCLUDING REMARKS

The following conclusions and final remarks may be made.

(a) The hydrodynamic and thermodynamic characteristics of confined axisymmetrical, gaseous turbulent diffusion flames can be predicted by the use of the k–ϵ turbulence model together with two additional mathematical models: one for the rate of chemical reaction and the other for the concentration fluctuations. The influences on the time-mean density of the fluctuations in both concentration and temperature can and should be accounted for. The

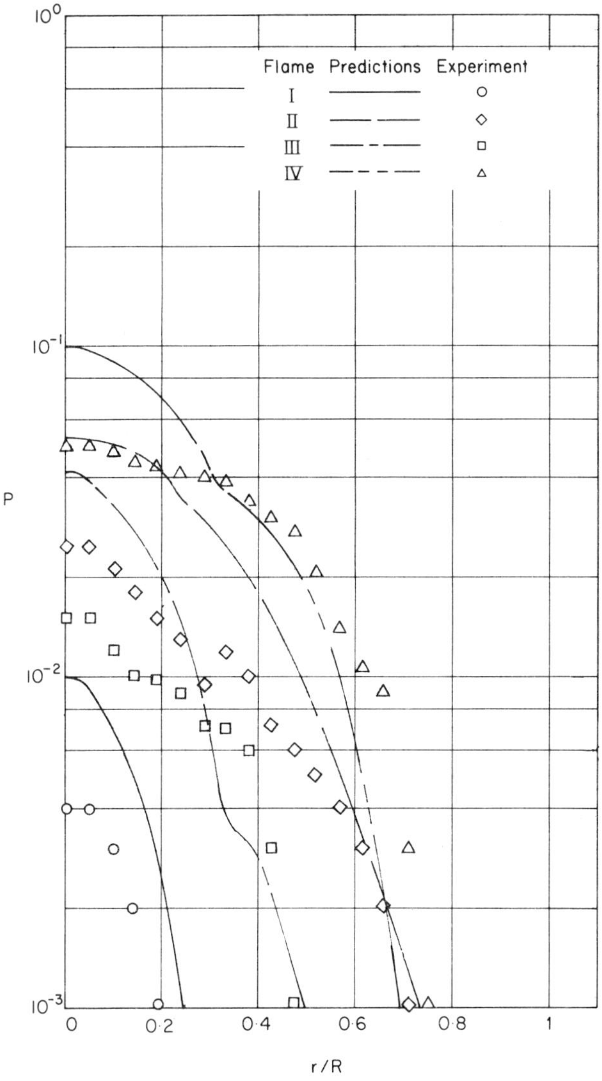

Fig. 12 (cont.) *d.* $z/D = 3.3$.

predictions made compare reasonably well with the few measurements that are available.

(b) A procedure has been described for calculating the probability of chemical reaction in turbulent diffusion flames. Its use in the present work has produced predictions which agree fairly well with measurements. Since the procedure makes use of local turbulence quantities only, there is reason to hope that the method can be applied more generally to other geometries.

(c) Since no direct measurements are available for concentration fluc-

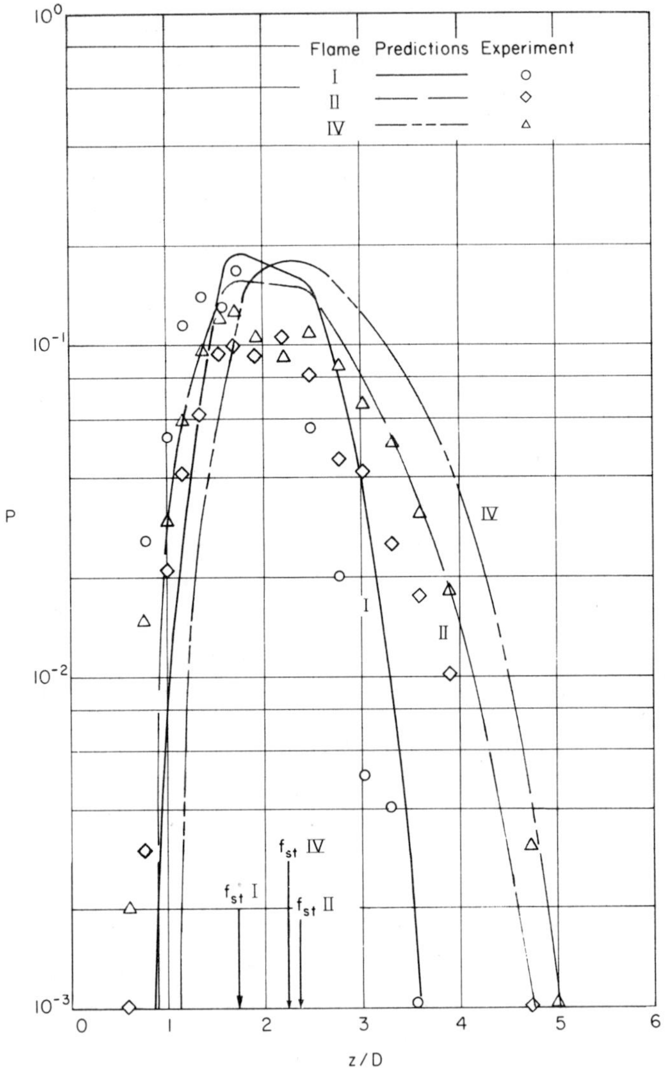

Fig. 13 Variation of **P** along axis of combustion chamber for flames I, II and IV.

tuations of stable species in turbulent flames, it is not possible to check the predictions of g. The good agreement between measurements and predictions of g in isothermal flows (Elghobashi *et al.* [45]) is the only available confirmation; one must, however, be cautious in extrapolating this conclusion to include turbulent flames until more evidence is available.

(d) The clipped Gaussian distribution proposed for the concentration fluctuations has proved to be realistic, in that its use leads to predicted mean values of the mass fractions and density all exhibiting the expected behaviour.

The predicted values of the time-mean temperatures are somewhat lower than available measurements for similar flows. This is thought to be mainly due to the predicted levels of temperature fluctuations being too high.

(e) The assumptions made in predicting the probability of reaction **P** and hence the extent of the flame brush, produce results which on the whole compare reasonably well with the measurements.

(f) The effect of chemical reaction on turbulence has so far been accounted for only through the spatial variation of the time-mean density. However, there are terms in the exact equations for k and g which represent the explicit influence of the chemical-reaction rate on both quantities (Bray [5], [46]). The modelling of these terms is required to assess their effects.

(g) In order to account for the heat transfer term from the combustion-chamber walls, two partial differential equations in a form similar to that of Eq. (2.3), are required, for the time-mean enthalpy, h, and its mean square fluctuations, $<h'^2>$. The restriction that a linear relationship exists between the enthalpy and mixture fraction may then be removed. Equations for the radiation fluxes would then have to be solved simultaneously.

(h) A need exists for measurements of the probability density function of stable-species concentration fluctuations in turbulent diffusion flames in order to assess the validity of the proposed function or its alternatives. This type of measurement requires instantaneous-response probes of high spatial resolution. Determination of this function for temperature fluctuations would be of great importance and might yield useful information about concentration fluctuations.

(i) The use of laser anemometry in measuring the time-mean and fluctuations of velocity seems to be very promising. A combination of these data with those of ionization probes would enable the structure of turbulent flames to be better understood.

REFERENCES

1. Hawthorne, W.R., Weddell, D.S. and Hottel, H.C. "Mixing and Combustion in Turbulent Gas Jets", *Third International Symposium on Combustion*, p.266, 1949.
2. Pompei, F. and Heywood, J.B. "The Role of Mixing in Burner Generated Carbon Monoxide and Nitric Oxide", *Combustion and Flame* 19, 407, 1972.
3. Bilger, R.W. and Kent, J.H. "Concentration Fluctuations in Turbulent Jet Diffusion Flames", Charles Kolling Research Laboratory, Technical Note F-46, University of Sydney, Australia, 1972.
4. Favre, A. "Equations des Gaz Turbulents Compressible", *J. de Méchanique* 4, 361, 1965.
5. Bray, K.N.C. "Equations of Turbulent Combustion – I. Fundamental Equations of Reacting Turbulent Flow", University of Southampton, Department of Aeronautics and Astronautics, Southampton, AASU Re. No. 330. 1973.
6. Elghobashi, S.E. "Characteristics of Gaseous Turbulent Diffusion Flames in Cylindrical Chambers", Ph.D. Thesis, University of London, 1974.
7. Spalding, D.B. "Mathematische Modelle turbulenter flammen", *VDI Berichte* 146, VDI Verlag, Dusseldorf, pp.25-30, 1970.
8. Kent, J.H. and Bilger, R.W. "Turbulent Diffusion Flames", Charles Kolling Research Laboratory Technical Note F-37, Department of Mechanical Engineering, The University of Sydney, Australia, 1972.

9. Wolfshtein, M. "Convection Processes in Turbulent Impinging Jets", Ph.D. Thesis, University of London, 1967.
10. Gosman, A.D., Pun, W.M., Runchal, A.K., Spalding, D.B. and Wolfshtein, M. "Heat and Mass Transfer in Recirculating Flows", Academic Press, 1969.
11. Mellor, G. and Herring, H.J. "A Survey of the Mean Turbulent Field Closure Models", *AIAA J.* **11**, 590, 1973.
12. Reynolds, W.C. "Computation of Turbulent Flows", *Ann. Rev. Fluid Mech.* **8**, 1976.
13. Launder, B.E., Morse, A., Rodi, W. and Spalding, D.B. "The Prediction of Free Shear Flows – A Comparison of the Performance of Six Turbulence Models", *Proc. NASA Langley Free Shear Flows Conf.* NASA SP 321, 1973.
14. Hinze, J.O. "Turbulence", McGraw-Hill, New York, 1959.
15. Davydov, B.I. "On the Statistical Dynamics of an Incompressible Turbulent Fluid", *Dok. Akad. Nauk. SSSR* **136**, pp.47-50, 1961. (*Soviet Physics – Doklady* **6**, p.10 (1961)).
16. Harlow, F.H. and Nakayama, P.I. "Transport of Turbulence Energy Decay Rate", Los Alamos Scientific Laboratory of the University of California, Report LA-3854, 1968.
17. Daly, B.J. and Harlow, F.H. "Transport Equations in Turbulence", *Phys. Fluids* **13**, 2634, 1970.
18. Hanjalić, K. and Launder, B.E. "A Reynolds Stress Model of Turbulence and Its Application to Asymmetric Shear Flows", *J. Fluid Mech.* **52**, 609, 1972.
19. Jones, W.P. and Launder, B.E. "The Prediction of Laminarization with a 2-equation Model of Turbulence", *Int. J. Heat & Mass Trans.* **15**, 301, 1972.
20. Günther, R. and Simon, H. "Turbulence Intensity, Spectral Density Functions, and Eulerian Scales of Emission in Turbulent Diffusion Flames", *Twelfth Symposium (International) on Combustion*, The Combustion Institute, pp.1069-1079, 1969.
21. Shvab, V.A. "Relation Between the Temperature and Velocity Fields of the Flame of a Gas Burner", Gos. Energ. izd., Moscow-Leningrad, 1948.
22. Zeldovich, Y.B. "On the Theory of Combustion of Initially Unmixed Gases", *Zhur. Tekhn. Fiz.* **19**, No.10, 1949. English Translation N.A.C.A. Tech. Mem. No. 1296 (1950).
23. Williams, F.A. "Combustion Theory", Chapters 1, 3, Addison-Wesley, Reading, Mass., 1965.
24. Glasstone, S. "Thermodynamics for Chemists", D. Von Nostrand Co., Inc., Princeton, New Jersey, 1946.
25. Tribus, M. and Desoto, S. "Thermodynamics", D. Van Nostrand Co., Inc., Princeton, New Jersey, 1961.
26. Csanady, G.T. "Concentration Fluctuations in Turbulent Diffusion", *J. Atmospheric Sci.* **24**, 21-28, 1967.
27. Spalding, D.B. "Concentration Fluctuations in a Round Turbulent Free Jet", *Chem. Eng. Sci.* **26**, 95-107, 1971.
28. Thring, M.W. and Newby, M.P. "Combustion Length of Enclosed Turbulent Jet Flames", *Fourth Symposium (International) on Combustion*, Williams and Wilkins, 789-796, 1953.
29. Richardson, J.M., Howard, H.C., Jr. and Smith, R.W., Jr. "The Relation Between Sampling-tube Measurements and Concentration Fluctuations in a Turbulent Gas Jet", *Fourth Symposium (International) on Combustion*, Williams and Wilkins, pp.814-817, 1953.
30. Rhodes, R.P. and Harsha, A.P.T. "On Putting the 'Turbulent' in Turbulent Reacting Flow", AIAA Paper No.72-68, New York, 1972.
31. Bush, W.B. and Fendell, F.E. "On Diffusion Flames in Turbulent Shear Flows", Project Squid Technical Report TRW-7-PU, Purdue University, Lafayette, Indiana, 1973.
32. Naguib, A.S., Private communication, Imperial College, 1973.
33. Bendat, J.S. and Piersol, A.G. "Random Data, Analysis Measurement Procedures", Wiley-Interscience, New York, p.66, 1971.
34. Launder, B.E. and Spalding, D.B. "The Numerical Computations of Turbulent Flows", *Computer Methods in Applied Mechanics and Engineering* **3**, 269-289, 1974.

35. El-Mahallawy, F.M., Lockwood, F.C. and Spalding D.B. "An Experimental and Theoretical Study of the Turbulent Mixing in a Cylindrical, Gas-fired Furnace", *Combustion Institute European Symposium*, Academic Press, pp.633-638, 1973.
36. Green, J.A. and Sugden, T.M. "Some Observations on the Mechanism of Ionization in Flames Containing Hydrocarbons", *Ninth Symposium (International) on Combustion*, pp.607-631, 1963.
37. Calcote, H.F. and King, I.R. "Studies on Ionization in Flames by Means of Langmuir Probes", *Fifth Symposium (International) on Combustion*, Reinhold, pp.423-434, 1955.
38. Batchelor, G.M. "Small-scale Variation of Convected Quantities Like Temperature in Turbulent Fluid", *J. Fluid Mech.* 5, 113, 1959.
39. Tennekes, H. and Lumley, J.L. "A First Course in Turbulence", MIT Press, 1972.
40. Hilst, G.R., Donaldson, C. du P., Teske, M., Contiliano, R. and Freiberg, J. "The Development and Preliminary Application of an Invariant Coupled Diffusion and Chemistry Model", A.R.A.P. Report No.193, prepared under Contract NAS1-11433 for NASA, 1973.
41. Elghobashi, S.E. and Pun, W.M. "A Theoretical and Experimental Study of Turbulent Diffusion Flames in Cylindrical Furnaces", Paper presented at the *Fifteenth Symposium (International) on Combustion*, Tokyo, Japan, 1974.
42. Lenze, B. "Turbulenzverhalten und Ungemischtheit von Strahlen und Strahlflammen", Ph.D. Thesis, Universität, Karlsruhe, 1971.
43. Günther, R. and Lenze, B. "Exchange Coefficients and Mathematical Models of Jet Diffusion Flames", *Fourteenth Symposium (International) on Combustion*, The Combustion Institute, pp.675-687, 1973.
44. Vranos, A., Faucher, J.E. and Curtis, W.E. "Turbulent Mass Transport and Rates of Reaction in a Confined Hydrogen-air Diffusion Flame", *Twelfth Symposium (International) on Combustion*, The Combustion Institute, pp.1051-1057, 1969.
45. Elghobashi, S.E., Pun, W.M. and Spalding, D.B. "Concentration Fluctuations in Isothermal Turbulent Confined Coaxial Jets", Imperial College Heat Transfer Report HTS/73/54 - reprinted as HTS/75/23, 1973.
46. Bray, K.N.C. "Kinetic Energy of Turbulence in Flames", University of Southampton, Department of Aeronautics and Astronautics, Southampton, AASU Rep. No.332, 1974.
47. Launder, B.E. and Spalding, D.B. "Mathematical Model of Turbulence", Academic Press, London, 1972.

NOMENCLATURE

c_p	specific heat
c_μ	constant in the turbulence model
E	activation energy
F	mean of the unclipped Gaussian distribution
f	time-mean mixture fraction
g	time-mean square of fluctuations of mixture fraction
H	lower value of the heat of reaction of fuel
h	time-mean enthalpy
i	stoichiometric mass of oxidant
k	kinetic energy of turbulence
ℓ	length scale of turbulence proportional to the size of the energy-containing eddies
M_j	molecular weight of species j

m_j time-mean mass fraction of species j
n_+ concentration of positive ions
P probability of chemical reaction
p absolute pressure
$P(\tilde{f})$ probability density function of the mixture fraction f
$\mathcal{R}$ universal gas constant
R radius of chamber
r distance in radial direction
T time-mean temperature
t time or time scale
u time-mean velocity in z-direction
U unit-step function
v time-mean velocity in radial direction
w time-mean square of vorticity fluctuations
z distance in axial direction

Greek Symbols

δ dirac delta function
ϵ kinematic dissipation rate of turbulence energy
η Kolmogorov's microscale of length, $\nu^{3/4}/\epsilon^{1/4}$
λ "reaction range" variable
μ molecular viscosity
μ_t turbulent viscosity
μ_{eff} effective viscosity, $\equiv \mu + \mu_t$
ν kinematic viscosity, $\equiv \mu/\rho$
ρ time-mean density
σ standard deviation of the unclipped Gaussian distribution
$\sigma_k, \sigma_\epsilon, \sigma_f, \sigma_g$ effective Schmidt number including both molecular and turbulent effects of subscripted entity
τ_s wall shear-stress
ψ stream function
ω vorticity

Subscripts

0 condition at the entering unpremixed oxidant stream
1 condition at the entering unpremixed fuel stream
℄ condition at the axis of symmetry
fu unburned fuel
ox unburned oxidant
pr products of combustion
st stoichiometric

Superscript

~ instantaneous value
′ fluctuating component
* dimensionless quantity

$$\underset{\text{instantaneous}}{(\tilde{\ })} \equiv \underset{\text{time-mean}}{(\ )} + \underset{\text{fluctuating component}}{(\ )'}$$

Special Notation

$\langle(\ )'^2\rangle^{1/2}$	R.M.S.
$\langle\ \rangle$	time-mean of a fluctuating component or correlation

CALCULATION OF CHEMICALLY REACTING FLOWS WITH COMPLEX CHEMISTRY

David T. Pratt
The University of Utah

ABSTRACT

This article describes a recently developed, efficient computational technique which allows the extension of any existing "cold flow" hydrodynamic code to include calculation of chemically complex, equilibrium or finite-rate chemistry. The success of the method follows from the use of two different field solution schemes, each optimal to the appropriate system of equations. The hydrodynamic field equations, which are weakly coupled and mildly non-linear, are best solved line-by-line for each variable of interest. The thermochemical field equations, on the other hand, are strongly coupled and highly non-linear, and are therefore best solved simultaneously, point-by-point. The two solution schemes are "super-iterated" until field values of the hydrodynamic and thermochemical variables are converged.

Unlike conventional chemical-kinetic integration schemes, the present technique is a zero-dimensional, rather than a one-dimensional formulation, so that it may be applied to flows of any space dimensionality, with or without time dependence. The technique is unconditionally stable with respect to the computational cell time constant, thus obviating the need for "time-splitting" in flows with differing time constants. In transient calculations this stability allows the use of large time steps, at the expense of accuracy. In steady flows, however, the need to time-integrate to the steady state is completely eliminated.

An example application is given, for turbulent combustion of initially unmixed methane and air in a shear layer.

1. INTRODUCTION

The computer modelling of chemically reacting flows has long been recognized as an extremely difficult problem because of the detail required to describe the interaction of many reacting chemical species with the fluctuating temperature and velocity fields which characterize a turbulent reacting flow.

This problem received much attention immediately after World War II, due largely to the interest in development of high-intensity continuous combustion devices such as turbojets, gas turbines and chemical rocket engines, and the need to describe propagation rates and other properties of detonation waves. With suitable descriptions of these phenomena in hand by the 1960's, and with receding public interest in aerospace technology in general, interest in this problem began to wane. At the same time, however, an increasing concern in combustion-generated air pollution, and more recently the need to predict combustion stability and pollutant-generation characteristics of potential alternate fuels for energy-conservation purposes, has rekindled interest in this problem.

By the end of the 1960's the accumulated literature on chemically reacting flows fell largely into two classes: (1) complex, homogeneous finite-rate chemistry in simple flows, and (2) finite-rate turbulent mixing in non-reacting or infinitely fast-reacting (chemically equilibrated) flows. The finite-rate chemistry literature has been authored mainly by physical chemists who by training have relatively little understanding of or interest in turbulence phenomena. Thus studies of chemical reaction rates in static reactors, flow reactors and shock tubes have all been conducted in systems in which precautions are taken so that mixing in any form is absent. The presence of field inhomogeneities due to axial diffusion or due to boundary-layer effects were regarded as aberrations of the desired homogeneous chemical field, and these effects were ignored altogether or (often clumsily) corrected for. One controversial exception to this approach has been the investigation of chemical-kinetic processes in well-stirred reactors, where the desired spatial homogeneity is approached, not by avoiding the occurrence of gradients, but rather by utilizing sonic turbulent jets to achieve such rapid mixing that the relatively slow chemical reaction rates causes the chemical composition of the field to be nearly homogeneous.

The second body of literature, which deals with turbulence *per se* and turbulent mixing in non-reacting or fully reacted flows, has been developed largely by fluid mechanicists with strong interest in mathematical and numerical-analysis techniques, and relatively little interest in the details of the molecular interactions that characterize chemical reaction. When combustion has been considered, it has frequently been represented by a single, global rate equation which explicitly relates a total energy release rate to local field values of the temperature and concentrations of one or at most two species.

By 1970, it had become apparent that our understanding of the mechanisms of pollutant formation in flames and furnaces, and of pollutant conversion and destruction in the atmosphere, required a balanced description of the interaction of both finite-rate chemistry and turbulent mixing. For example, in the case of the proposed use of "low-BTU" fuels, it was no longer sufficient merely to characterize the heat release rate of the fuel *when* it was burned; it became necessary to know under what conditions of fuel-air mixture ratio and turbulent mixing one fuel could be made to burn at all. In addition, the rates of dispersion and chemical reaction of air pollutants in the atmosphere are now known to be controlled by *both* the finite rates of mixing and of chemical reaction. Unfortunately, the state of our knowledge of both fundamental chemical kinetics and of turbulent mixing phenomena is not yet equal to the new task, and our present inability to describe the coupled effects of these finite-rate phenomena is a serious impediment to the formulation of rational public policies for control of combustion-generated air pollutants.

The articles in this volume by Professor Libby and Dr. Elghobashi deal with modern developments in the treatment of coupled reaction and mixing phenomena in convective flows, primarily from the point of view of fluid mechanics. The present article, on the other hand, is written primarily from a background of experience with homogeneous finite-rate chemistry. The purpose of the present article is to illustrate a recently developed, efficient computational technique which enables the inclusion of complex, finite-rate chemistry into any existing solution scheme for the flow field. While a mechanism for modelling of inhomogeneous reacting flow fields (i.e., accounting for the coupled effects of fluctuating velocities and thermochemical variables) is included in this formulation, its viability had not been demonstrated at the time this article was written.

2. GENERAL DESCRIPTION OF THE METHOD

2.1 *Rationale*

D.B. Spalding and colleagues at Imperial College [1,2] have shown that finite-difference approximations to the differential equations for conservation of mass, momentum, energy, or any variable of interest ϕ, may be expressed in a standard form

$$\sum_d A_d\phi_P = \sum_d A_d\phi_d + S_\phi \tag{2.1}$$

where the subscript d ("direction") refers to the six adjacent nodes as illustrated in Fig.1, namely N, S, E, W, H and L, and to PP, which denotes the point P at a previous time step. The A_d's are convection and diffusion coefficients, except A_{PP}, which is the influence coefficient for conditions at point P at a previous time step. The exact forms of the A_d's and A_{PP} depend on the nature of the governing equations (elliptic, parabolic or hyperbolic; time-dependent or steady) and on the form of differencing approximation used (upwind or donor cell, central, etc.) [3].

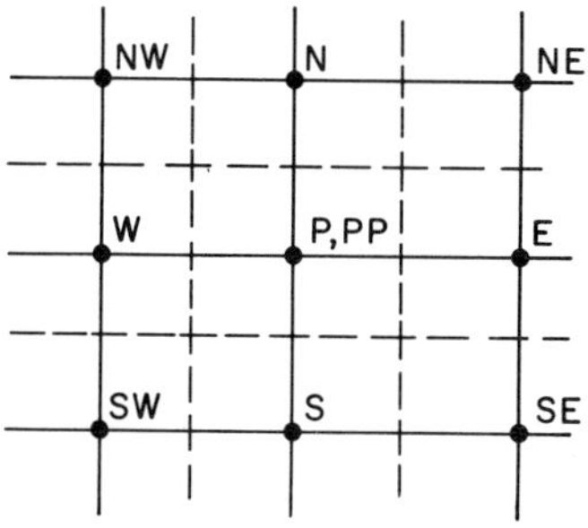

Fig.1 Conventional staggered finite-difference grid for calculation within a flow domain [1,2]. Points H ("high") and L ("low") are above and below the plane of E–W–N–S–P. PP refers to conditions at P at the previous time step.

The "source" term S_ϕ in Eq. (2.1) is frequently linearized,

$$S_\phi = B + C\phi_P , \tag{2.2}$$

and the resulting equations for any variable ϕ solved either point-by-point by Gauss-Seidel (GS) iteration [3], with Eq. (2.1) in the form

$$\phi_P = \frac{\sum_d A_d \phi_d + B}{(\sum_d A_d - C)} \tag{2.3}$$

or line-by-line, by means of the tri-diagonal matrix algorithm (TDMA) [3], with Eq. (2.1) in the form

$$\phi_P = \frac{\sum_{d \neq N,S} A_d \phi_d + B}{(\sum_d A_d - C)} + \frac{A_N \phi_N}{(\sum_d A_d - C)} + \frac{A_S \phi_S}{(\sum_d A_d - C)} \tag{2.4}$$

These two practices have been successful when applied to the solution of conservation equations in which the source terms are linear or mildly non-linear. As Dr. Elghobashi has commented in the preceding article, TDMA is generally faster than GS iteration, because there is simultaneous solution for many values of ϕ, while GS updates only one value of ϕ at a time. In physical terms, the linearization of Eq. (2.2) and solution by GS or TDMA give satisfactory rates of convergence when local variations in ϕ_P depend more strongly on the values of the *same* variable at neighbouring nodes than on the values of *other* variables at the same point P.

However, for conservation of chemical species in a reacting flow, the chemical-kinetic source term for mole numbers of species i is

$$S_{\sigma_i} = - \sum_{j=1}^{JJ} (a'_{ij} - a''_{ij})(R_j - R_{-j}) , \tag{2.5}$$

where R_j and R_{-j} are the rates of forward and reverse reactions j, respectively; other symbols are defined in the Nomenclature. The forward reaction rate R_j may be expressed in a modified Arrhenius form,

$$R_j = X_j \, 10^{B_j} \, T^{N_j} \exp(-T_j/T)(\rho\sigma_m)^{\overline{a_j}} \prod_{k=1}^{NS} (\rho\sigma_k)^{a'_{kj}} , \tag{2.6}$$

where X_j is the *contact index* for reaction j [4,5], of which more will be said in Section 4.

It may be supposed from inspection of Eqs (2.5) and (2.6) that the species-i

mole number at point P, $(\sigma_{i,P})$, may depend more strongly on (any or) *all* of the mole numbers and the temperature at point P than on the values of $\sigma_{i,d}$ at adjacent nodes. If this is indeed true then two principles for solving σ_i equations may be postulated:

(1) Solutions of the equations for $(\sigma_i,\ i = 1, NS)$ should be simultaneous, point-by-point, and coupled with T through some form of the energy equation.

(2) Due to the non-linearities of the rate expressions, Eqs (2.6), the use of derivative information may be valuable to achieve rapid convergence when large local variations in $(\sigma_i,\ i = 1, NS)$ or in T occur.

2.2 *Scheme for Solution of Species Equations*

The above principles are satisfied by the following formulation:

With the mole number of species i, σ_i, substituted for ϕ, Eq. (2.1) can be rewritten

$$\sum_d A_d \left[\sigma_{i,P} - \frac{\sum_d A_d \sigma_{i,d}}{\sum_d A_d} \right] = S_{\sigma_i} \quad , \tag{2.7}$$

where S_{σ_i} is necessarily given by the highly non-linear Eq. (2.5) rather than by the linear relation, Eq. (2.2).

Equations (2.7) are NS in number, where NS is the number of chemical species being considered, and are coupled through the σ_i source terms, Eqs (2.5). The thermal energy equation is also required to determine the temperature T, also in the form of Eq. (2.7):

$$\sum_d A_d \left[H_p - \frac{\sum_d A_d H_d}{\sum_d A_d} \right] = S_H \quad , \tag{2.8}$$

where $H \equiv \sum_{i=1}^{NS} h_i \sigma_i$ is the sensible-plus-chemical static enthalpy, and where S_H is the radiative heat transfer and kinetic heating "enthalpy source" term, which may be expressed with sufficient generality by any differentiable function of T.

With the definition and substitution of new variables

$$A_P \equiv \sum_d A_d \quad , \tag{2.9}$$

$$\sigma_{i,P}^* \equiv \frac{\sum_d A_d \sigma_{i,d}}{A_P} \quad , \tag{2.10}$$

and

$$H_P^* \equiv \frac{\sum_d A_d H_d}{A_P} , \tag{2.11}$$

equations (2.7) and (2.8) may be written in the form

$$A_P(\sigma_{i,P} - \sigma_{i,P}^*) = S_{\sigma_i} , \qquad i = 1, NS \tag{2.12}$$

and

$$A_P(H_P - H_P^*) = S_H , \tag{2.13}$$

respectively.

Rewriting Eqs (2.12) in the functional form

$$f_i = A_P(\sigma_{i,p} - \sigma_{i,P}^*) - S_{\sigma_i} , \tag{2.14}$$

a set of Newton-Raphson correction equations may be written,

$$\sum_{k=1}^{NS} \left(\frac{\partial f_i}{\partial \sigma_k}\right)^{(o)} \Delta\sigma_k^{(o)} = -f_i^{(o)} , \qquad i = 1, NS \tag{2.15}$$

where $(\sigma_i^{(o)}, i = 1, NS)$ and $T^{(o)}$ is the initial estimate set.

Equations (2.15) may be solved iteratively, together with a similar correction equation for the temperature from Eq. (2.13), until the corrections $(\Delta\sigma_k = 1, NS)$ and ΔT vanish, or nearly so.

The overall strategy proposed is to solve the hydrodynamic field equations for conservation of mass and momentum (which have weak source terms) by means of either GS or TDMA iteration, with the equations in the standard forms Eqs (2.3) or (2.4). With the convection and diffusion coefficients A_d's thus determined (modified by suitable Prandtl or Schmidt numbers if desired), Eqs (2.7) and (2.8) in the form of Eq. (2.15) may be solved iteratively by pivotal Gaussian elimination, point-by-point, for simultaneous determination of all the thermochemical variables $(\sigma_i, i = 1, NS)$ and T. The new values of mass density thus determined are used as data for a new GS or TDMA sweep to determine velocities and pressure, and the cycle of "super-iteration" between the hydrodynamic and thermochemical fields repeated until a stationary solution is obtained.

The initial estimate set for Eqs (2.15) may be obtained either from the solution values from previous iterations, or from chemical equilibrium states calculated by minimization of the mixture Gibbs function at assigned values of pressure and enthalpy.

Further elaboration of this solution scheme is given in subsequent sections.

3. CALCULATION OF CHEMICAL EQUILIBRIUM

3.1 General

A numerically efficient scheme for calculation of chemically complex equilibrium distributions of species mole numbers is of value in its own right, in cases such as stoichiometric hydrogen-oxygen combustion where the chemical kinetic rates are known to be so fast compared to turbulent mixing rates that they may be taken to be infinite to good approximation. In addition, chemical equilibrium solutions are necessary as initial estimates for the iterative solution of finite-rate non-equilibrium states.

Three schemes for determination of equilibrium states at prescribed temperature and pressure may be considered, namely: Gibbs function minimization, equilibrium constant formulation, and equating of forward and reverse rates. The first two schemes are well known to be equivalent, "static equilibrium" formulations based on the Gibbs function minimization principle for a mixture at specified temperature and pressure. The third, "dynamic equilibrium" scheme, is not of practical use, as it is essentially a limiting case of finite-rate kinetics based on infinite reaction time, and therefore has all of the difficulties associated with finite-rate chemistry solutions.

Of the two static equilibrium schemes, the equilibrium constant approach is the most familiar and is most easily formulated. In this approach, where *NS* equilibrium mole numbers are sought, *NLM* equations for conservation of atomic species are formulated, and the remaining (*NS–NLM*) required equations are formulated by postulating (*NS–NLM*) stoichiometrically independent elementary reactions, each of which has associated with it an "equation of reaction equilibrium" involving a temperature-dependent equilibrium "constant". Unfortunately, a scheme to solve any arbitrarily chosen set of these *NS* nonlinear algebraic equations does not exist. In addition, different sets of reactions are required as fuel-air equivalence ratios are varied from lean to rich [6,7].

A vastly superior computational scheme is the Gibbs function minimization approach, which does not require a choice of individual reactions, and has been formulated by Gordon and McBride [8] in such a way that only *NLM* equations and unknowns are required, rather than *NS*. Further, a foolproof scheme for solution for any given reactants distribution has been devised. This formulation, adapted only slightly for present purposes from Ref. [8], will be developed below.

3.2 Gibbs Function Minimization Formulation

For a mixture of reacting gases at prescribed temperature and pressure, chemical equilibrium obtains when the Gibbs function of the mixture is a minimum, subject to conservation of atomic species as specified in the set of initial reactant mole numbers. If the mixture enthalpy is specified instead of the temperature, an equation of conservation of thermal energy is required as an additional constraint.

For a mixture of ideal gases, the partial molar Gibbs function is given by

$$g_k \equiv h_k - Ts_k = h_k^o - Ts_k^o + RT \log \frac{\sigma_k}{\sigma_m} + RT \log \frac{P}{P_o} \tag{3.1}$$

where P_o is standard atmospheric pressure, and where other symbols are defined in the Nomenclature. For an ideal gas, Eq. (3.1) is also equal to the chemical potential for species k in the mixture.

The mass-specific Gibbs function for the mixture is given by

$$G \equiv \sum_{k=1}^{NS} \sigma_k g_k \ , \tag{3.2}$$

and the requirement for chemical equilibrium is

$$dG = \sum_{j=1}^{NS} \frac{\partial G}{\partial \sigma_j} d\sigma_j = 0 \ , \qquad (d^2 G > 0) \ , \tag{3.3}$$

subject to given P and T, and conservation of atomic species

$$\sum_{k=1}^{NS} a_{ik}^L \sigma_k - b_i^* = 0 \ . \qquad i = 1, NLM \ . \tag{3.4}$$

Subsequently, the symbol b_i will be used for the first term on the left-hand side of Eq. (3.4).

As indicated above, when the mixture enthalpy H^* is specified, rather than T, the thermal energy equation is also needed:

$$H^* - \sum_{k=1}^{NS} h_k \sigma_k - (Q/A_p) = 0 \ , \tag{3.5}$$

where H^* is defined by Eq. (2.11), and

$$Q \equiv -S_H \ , \tag{3.6}$$

S_H having been defined by Eq. (2.8) and A_P by Eq. (2.9).

The Lagrange method of undetermined multipliers is now employed, as follows:

1. Write out Eq. (3.3), with Eqs (3.1) and (3.2) incorporated:

$$dG = \sum_{j=1}^{NS} \frac{\partial}{\partial \sigma_j} \left\{ \sum_{k=1}^{NS} \sigma_k \left[g_k^o + RT \log \sigma_k - RT \log \left(\sum_{i=1}^{NS} \sigma_i \right) + RT \log \frac{P}{P_o} \right] \right\} d\sigma_j$$

$$= \sum_{j=1}^{NS} \left\{ \sum_{k=1}^{NS} \left[\sigma_k \left(RT\,\sigma_k^{-1}\,\delta_{jk} - RT\,\sigma_m^{-1} \right) + g_k\,\delta_{jk} \right] \right\} d\sigma_j = 0 \,, \tag{3.7}$$

where δ_{jk} is the Kronecker delta.

Simplifying Eq. (3.7),

$$dG = \sum_{k=1}^{NS} g_k \, d\sigma_k = 0 \,. \tag{3.8}$$

2. Take the differential of each constraint equation, Eqs (3.4), and multiply by an undetermined constant multiplier λ_i:

$$\lambda_i \sum_{k=1}^{NS} a_{ik}^L \, d\sigma_k = 0 \,, \qquad i = 1, NLM \tag{3.9}$$

3. Add Eqs (3.8) and (3.9), and combine coefficients of $d\sigma_k$:

$$\sum_{k=1}^{NS} \left[g_k + \sum_{i=1}^{NLM} \lambda_i \, a_{ik}^L \right] d\sigma_k = 0 \,. \tag{3.10}$$

The coefficients of the $d\sigma_k$'s in Eq. (3.10) are required to vanish independently; i.e., for arbitrary variations in $\{\sigma_k\}$ about the equilibrium values $\{\sigma_k{}^{eq}\}$. Therefore, a set of NS equations must be satisfied,

$$g_k + \sum_{i=1}^{NLM} \lambda_i \, a_{ik}^L = 0 \,, \qquad k = 1, NS \,, \tag{3.11}$$

in addition to the NLM constraint Eqs (3.4).

At this point, there are $(NS + NLM)$ equations for the $(NS + NLM)$ unknowns $\{\sigma_k{}^{eq},\ k = 1, NS\}$ and $\{\lambda_i,\ i = 1, NLM\}$, which is a greater number than the NS equations and unknowns required in the equilibrium constant formulation. The following section illustrates how the problem is reduced to an (NLM)-dimensional system of equations and unknowns.

3.3 Newton-Raphson Solution of the Extremum Equations

Defining non-dimensional Lagrange multipliers $\pi_i \equiv -\lambda_i / RT$, Eqs (3.11) and (3.4) may be expressed as functionals:

$$f_k \equiv \frac{g_k}{RT} - \sum_{i=1}^{NLM} \pi_i a_{ik}^L , \qquad k = 1, NS \tag{3.12}$$

and

$$f_i \equiv b_i - b_i^* , \qquad i = 1, NLM . \tag{3.13}$$

These functionals must each vanish at the equilibrium solution $\{ \sigma_k = \sigma_k^{eq} ,$ $k = NS \}$.

In addition, a functional

$$f_T \equiv -\frac{H^*}{RT} + \sum_{k=1}^{NS} \frac{h_k}{RT} \sigma_k + \frac{Q/A_p}{RT} \tag{3.14}$$

must be defined from Eq (3.5), if the mixture enthalpy H^* and specific heat loss Q are specified instead of T.

A system of Newton-Raphson correction equations for correction variables Δx_j are defined by

$$\sum_{j=1}^{N} \frac{\partial f_i}{\partial x_j} \Delta x_j = -f_i , \qquad i = 1, N \tag{3.15}$$

for an N-dimensional system of equations f_i and unknowns x_j. Following Gordon and McBride [8], a judicious choice of correction variables in the present case is $(\Delta \log \sigma_k, \ k = 1, NS)$, $\Delta \log \sigma_m$ and $\Delta \log T$. The selection of $\Delta \log \sigma_m$ as an independent correction variable requires yet another functional to vanish at equilibrium, namely

$$f_M \equiv \sum_{k=1}^{NS} \sigma_k - \sigma_m . \tag{3.16}$$

With this set of correction variables, and Eqs (3.12) substituted into Eq. (3.15), there results for conservation of the jth species,

$$\sum_{k=1}^{NS} \frac{\partial f_j}{\partial \log \sigma_k} \Delta \log \sigma_k + \frac{\partial f_j}{\partial \log \sigma_m} \Delta \log \sigma_m$$

$$+ \frac{\partial f_j}{\partial \log T} \Delta \log T = -f_j , \quad j = 1, NS \tag{3.17}$$

Writing out the species-j functional Eq. (3.12) in full,

$$f_j \equiv \frac{g_j^o}{RT} + \log \sigma_j - \log \sigma_m + \log (P/P_o) - \sum_{i=1}^{NLM} \pi_i a_{ij}^L \,, \qquad j = 1, NS \,, \tag{3.18}$$

the motivation for choosing $\Delta \log \sigma_m$ as an independent correction variable is now apparent. Noting that

$$g_j^o \equiv h_{f_{298_j}}^o + \int_{298}^{T} C_{p_j} dT' - T \left[s_{298_j}^o + \int_{298}^{T} C_{p_j} \frac{dT'}{T'} \right] , \tag{3.19}$$

the partial derivative coefficients in Eq. (3.17) may be expressed fully as

$$\frac{\partial f_j}{\partial \log \sigma_k} = \delta_{jk} \,, \qquad k = 1, NS \,, \tag{3.20}$$

$$\frac{\partial f_j}{\partial \log \sigma_m} = -1 \,, \tag{3.21}$$

and

$$\frac{\partial f_j}{\partial \log T} = T \frac{\partial f_j}{\partial T} = - \frac{h_j}{RT} \,. \tag{3.22}$$

Substitution of Eqs (3.20) – (3.22) into Eqs (3.17) yields, for species k,

$$\Delta \log \sigma_k - \Delta \log \sigma_m - \frac{h_k}{RT} \Delta \log T = - \frac{g_k}{RT} + \sum_{i=1}^{NLM} \pi_i a_{ik}^L \,, \qquad k = 1, NS \,. \tag{3.23}$$

The *NLM* atom-conservation correction equations are obtained from the functionals Eqs (3.13):

$$f_i \equiv \sum_{j=1}^{NS} \sigma_j a_{ij}^L - b_i^* \,, \qquad i = 1, NLM \,. \tag{3.13}$$

The appropriate partial derivatives for the corresponding atom-i correction equations are

$$\frac{\partial f_i}{\partial \log \sigma_k} = \sigma_k a_{ik}^L \,, \tag{3.24}$$

$$\frac{\partial f_i}{\partial \log \sigma_m} = 0 \ , \tag{3.25}$$

and

$$\frac{\partial f_i}{\partial \log T} = 0 \tag{3.26}$$

The correction equations for atom-i conservation are therefore

$$\sum_{k=1}^{NS} \sigma_k\, a_{ik}^L\, \Delta \log \sigma_k = - \sum_{k=1}^{NS} \sigma_k\, a_{ik}^L + b_i^* \ , \quad i = 1,\ NLM \tag{3.27}$$

Because of the choice of $\Delta \log \sigma_m$ as an independent variable, it is now possible to eliminate $(\Delta \log \sigma_k,\ k = 1, NS)$ as independent correction variables by substitution of Eqs (3.23) into the NLM atom-j correction equations, Eqs (3.27):

$$\sum_{k=1}^{NS} \sigma_k\, a_{jk}^L \left\{ \Delta \log \sigma_m + \frac{h_k}{RT} \Delta \log T - \frac{g_k}{RT} \right.$$

$$\left. + \sum_{i=1}^{NLM} \pi_i\, a_{ik}^L \right\} = b_j^* - \sum_{k=1}^{NS} \sigma_k\, a_{jk}^L \ , \quad j = 1,\ NLM \ . \tag{3.28}$$

Rearranging terms, Eq. (3.28) may be rewritten as

$$\sum_{i=1}^{NLM} \left[\sum_{k=1}^{NS} a_{ik}^L\, a_{jk}^L\, \sigma_k \right] \pi_i + \sum_{k=1}^{NS} a_{jk}^L\, \sigma_k \quad \Delta \log \sigma_m +$$

$$\left[\sum_{k=1}^{NS} a_{jk}^L\, \sigma_k \frac{h_k}{RT} \right] \Delta \log T = (b_j^* - b_j) +$$

$$\sum_{k=1}^{NS} a_{jk}^L\, \sigma_k \frac{g_k}{RT} \ , \quad j = 1,\ NLM \ . \tag{3.29}$$

The correction equation for reciprocal mixture molecular weight σ_m is obtained by first taking partial derivatives of Eq. (3.16) with respect to the correction variables:

$$\frac{\partial f_m}{\partial \log \sigma_j} = \sigma_j \frac{\partial f_M}{\partial \sigma_j} = \sigma_j \sum_{k=1}^{NS} \frac{\partial \sigma_k}{\partial \sigma_j} = \sigma_j \ , \quad j = 1,\ NS \tag{3.30}$$

and

$$\frac{\partial f_M}{\partial \log \sigma_m} = \sigma_m \frac{\partial f_M}{\partial \sigma_m} = \sigma_m(-1) = -\sigma_m \tag{3.31}$$

and

$$\frac{\partial f_M}{\partial \log T} = 0\ . \tag{3.32}$$

Substitution of Eq. (3.16) and Eqs (3.30) – (3.32) into Eq. (3.15) yields the correction equation for reciprocal mixture molecular weight, σ_m:

$$\sum_{k=1}^{NS} \sigma_k\, \Delta \log \sigma_k - \sigma_m\, \Delta \log \sigma_m = \sigma_m - \sum_{k=1}^{NS} \sigma_k\ . \tag{3.33}$$

As before, $\Delta \log \sigma_k$ is eliminated by means of substitution from Eq. (3.23):

$$\sum_{k=1}^{NS} \sigma_k \left\{ \Delta \log \sigma_m + \frac{h_k}{RT}\, \Delta \log T - \frac{g_k}{RT} + \sum_{i=1}^{NLM} \pi_i\, a_{ik}^L \right\}$$

$$- \sigma_m\, \Delta \log \sigma_m = \sigma_m - \sum_{k=1}^{NS} \sigma_k\ . \tag{3.34}$$

Rearrangement of terms results in the σ_m-correction equation:

$$\sum_{i=1}^{NLM} \left[\sum_{k=1}^{NS} a_{ik}^L \sigma_k \right] \pi_i + \left[\sum_{k=1}^{NS} \sigma_k - \sigma_m \right] \Delta \log \sigma_m +$$

$$\left[\sum_{k=1}^{NS} \sigma_k \frac{h_k}{RT} \right] \Delta \log T = \sigma_m - \sum_{k=1}^{NS} \sigma_k + \sum_{k=1}^{NS} \sigma_k \frac{g_k}{RT}\ . \tag{3.35}$$

The correction equation for temperature is obtained in a similar manner: first, take the partial derivatives of Eq. (3.14) with respect to $(\log \sigma_k, k = 1, NS)$, $\log\ \sigma_m$ and $\log T$. Second, substitute these derivatives and Eq. (3.14) into Eq. (3.15); third, eliminate the terms $(\Delta \log \sigma_k,\ k = 1, NS)$ by substitution of Eq. (3.23) to yield the temperature correction equation:

$$\sum_{i=1}^{NLM} \left[\sum_{k=1}^{NS} a_{ik}^L \sigma_k \frac{h_k}{RT} \right] \pi_i + \left[\sum_{k=1}^{NS} \sigma_k \frac{h_k}{RT} \right] \Delta \log \sigma_m + \tag{3.36}$$

$$\left[\sum_{k=1}^{NS} \sigma_k \frac{C_{P_k}}{R} + \sum_{k=1}^{NS} \sigma_k \left(\frac{h_k}{RT}\right)^2 + \frac{1}{RT}\frac{\partial(Q/A_P)}{\partial \log T}\right] \Delta \log T = \sum_{k=1}^{NS} \sigma_k^* \frac{h_k^*}{RT}$$

$$- \sum_{k=1}^{NS} \sigma_k \frac{h_k}{RT} - \frac{(Q/A_P)}{RT} + \sum_{k=1}^{NS} \sigma_k \frac{h_k}{RT}\frac{g_k}{RT} \quad . \qquad (3.36) \text{ (cont.)}$$

3.3 Solution of the Correction Equations

By judicious selection of correction variables, the number of independent variables was reduced from $(NS + NLM)$ to $(NLM + 2)$, with a corresponding number of linear algebraic equations, Eqs (3.29), (3.35) and (3.36). Given an estimate set $(\sigma_k{}^{(o)},\ k = 1, NS)$, $\sigma_m{}^{(o)}$ and $T^{(o)}$, values for $(\pi_i,\ i = 1, NLM)$, $\Delta \log \sigma_m$ and $\Delta \log T$ are determined by solving the system of linear equations by pivotal Gaussian elimination. Substitution into Eq. (3.23) yields the corresponding corrections $(\Delta \log \sigma_k,\ k = 1, NS)$. New values of $(\sigma_k,\ k = 1, NS)$, σ_m and T arc then obtained by the relation

$$\sigma_k = \sigma_k^{(o)} + \eta \exp(\Delta \log \sigma_k) \ , \quad \text{etc.} \qquad (3.37)$$

where η is an underrelaxation or acceleration parameter.

Iteration is continued until absolute values of all of the corrections are less than some suitably small number ϵ. It may be of interest to note that the values of the Lagrange multipliers π_i do not appear explicitly. They are of no interest physically, and their current values are merely used in Eq. (3.23) for the species-k corrections, $\Delta \log \sigma_k$.

3.4 Control of the Underrelaxation Parameter

Gordon and McBride [8] have devised an empirical, "self-adjusting" underrelaxation parameter, which is the key to a rapidly converging solution even from very poor initial estimates. This scheme may be summarized as follows:

(i) For T and σ_m, and for σ_i for those species for which $(\sigma_i/\sigma_m) > 10^{-8}$ and for which $\Delta \log \sigma_i > 0$, a number η_1 is defined as

$$\eta_1 = \frac{2}{\max(|\Delta \log T|,\ |\Delta \log \sigma_m|,\ |\Delta \log \sigma_i|)} \ , \quad i = 1,\ NS \ . \qquad (3.38)$$

This causes the corrections which appear in the denominator of Eq. (3.38) to be scaled so that the corresponding variable (i.e., T, σ_m or σ_i for major species with positive corrections) will not be increased by more than $\exp(2) = 7.3891$ on any one iteration.

(ii) For species for which $(\sigma_i/\sigma_m) < 10^{-8}$ and for which $\Delta \log \sigma_i > 0$, a number η_2 is defined as

$$\eta_2 = \frac{\log(10^{-4}) - \log(\sigma_i/\sigma_m)}{\Delta \log \sigma_i \ \Delta \log \sigma_m} , \qquad i = 1, NS \tag{3.39}$$

This causes the mole number for species with mole fraction initially less than 10^{-8} to increase to no more than 10^{-4}.

(iii) The final choice of underrelaxation parameter η is defined by

$$\eta = \min(1, \eta_1, \eta_2) . \tag{3.40}$$

A new value of η is calculated at each iteration and applied to Eq. (3.37). Thus whenever current values of estimates are far from the solution point, η will be less than unity, and the iterative move toward the solution will be damped accordingly. Near the solution point, η will equal 1 and the final iterations will be undamped. No attempt is made to accelerate the solution; η is never greater than 1.

3.5 Convergence Criterion ϵ

The log mole number corrections $(\Delta \log \sigma_i,\ i = 1, NS)$ are all required to be less than $\epsilon = 0.01$.

Since, for small increments,

$$\Delta \log \sigma_i \approx \frac{\Delta \sigma_i}{\sigma_i} , \tag{3.41}$$

this criterion is equivalent to requiring convergence to within one percent for all mole numbers.

3.6 Initial Estimates

Because of the "self-adjusting" underrelaxation parameter described in Sec. 3.4, a "garbage" estimate set may be employed, again following Gordon and McBride [8]. For prescribed pressure and mixture enthalpy, the temperature is initially estimated at 3800°K, and the mole numbers $(\sigma_i,\ i = 1, NS)$ are all simply set equal to $(0.1/NS)$.

Remarkably, Gordon and McBride [8] claim, and this writer's experience confirms, that a converged solution is *always* obtained from the "garbage" estimates, and usually in ten or fewer iterations.

4. CALCULATION OF CHEMICAL-KINETIC STATIONARY STATES

4.1 Newton-Raphson Iteration Equations

The same correction variables, $(\Delta \log \sigma_i,\ k = 1,\ NS)$, $\Delta \log \sigma_m$ and $\Delta \log T$, are employed as were used for the Gibbs-function minimization equations, described in Section 3.

The functionals for the species-i conservation equations are given by Eqs (1.5) and (1.14):

$$f_i = A_P(\sigma_i - \sigma_i^*) + \sum_{j=1}^{JJ} (a'_{ij} - a''_{ij})(R_j - R_{-j}) \quad , \quad i = 1, NS \tag{4.1}$$

where the forward and reverse rates R_j and R_{-j} are given by

$$R_j = X_j\, 10^{B_j}\, T^{N_j} \exp\left[-\frac{T_j}{T}\right] (\rho\sigma_m)^{\overline{a}_j} \prod_{k=1}^{NS} (\rho\sigma_k)^{a'_{kj}} \quad , \tag{4.2}$$

and

$$R_{-j} = X_{-j}\, 10^{B_{-j}}\, T^{N_{-j}} \exp\left[-\frac{T_{-j}}{T}\right] (\rho\sigma_m)^{\overline{a}_j} \prod_{k=1}^{NS} (\rho\sigma_k)^{a''_{kj}} \ . \tag{4.3}$$

The contact indices X_j and X_{-j} are taken to be constants in this formulation. In fact, X_j and X_{-j} are more likely to be functionally dependent on the local values of (grad σ_i, $i = 1$, NS) and (grad T) than on the local values of $\{\sigma_i, i = NS\}$ and T [4,5]. It is assumed that, in an iterative solution for flow field values of (σ_i, $i = 1$, NS) and T, *explicit* finite-difference expressions for (grad σ_i) and (grad T) will be employed; that is, (grad $\sigma_i)_P$ and (grad $T)_P$ will be expressed in terms of differences in neighbour-node, "old" values of σ_{id} and T_d, and of node distances determined by the particular computational grid employed. This practice is necessary and consistent with the "zero-dimensionality" of the present formulation.

The correction equations for species-i mole numbers are

$$\sum_{k=1}^{NS} \frac{\partial f_i}{\partial \log \sigma_k} \Delta \log \sigma_k + \frac{\partial f_i}{\partial \log \sigma_m} \Delta \log \sigma_m + \frac{\partial f_i}{\partial \log T} \Delta \log T = -f_i \ , \qquad i = 1, NS \ . \tag{4.4}$$

The partial derivative coefficients of the correction variables in Eq. (4.4) are

$$\frac{\partial f_i}{\partial \log \sigma_k} = A_P\, \sigma_i\, \delta_{ik} + \sum_{j=1}^{JJ} (a'_{ij} - a''_{ij})(R_j a'_{kj} - R_{-j} a''_{kj}) \ , \tag{4.5}$$

$$\frac{\partial f_i}{\partial \log \sigma_m} = \sum_{j=1}^{JJ} (a'_{ij} - a''_{ij})(R_{-j} n''_j - R_j n'_j) \ , \tag{4.6}$$

and

$$\frac{\partial f_i}{\partial \log T} = \sum_{j=1}^{JJ} (a'_{ij} - a''_{ij}) \left[R_j \left\{ N_j + \frac{T_j}{T} - \overline{a}_j - n'_j \right\} \right.$$

$$- R_{-j}\left\{N_{-j} + \frac{T_{-j}}{T} - \overline{a}_j - n_j''\right\}\Bigg] \quad . \tag{4.7}$$

Alternative treatments of the rate expressions, Eqs (4.2) and (4.3) are possible. For example, if it is desired to construct the reverse rate data from the relation

$$k_{-j} = \frac{k_j [RT]^{(n_j'' - n_j')}}{K_j^P} \quad , \tag{4.8}$$

where K_j^P is the equilibrium constant for reaction j,

$$K_j^P \equiv \exp\left[\sum_{i=1}^{NS} (a_{ij} - a_{ij}'') \frac{g_i^o}{RT}\right] \quad , \tag{4.9}$$

then the reverse rate R_{-j} is given by

$$R_{-j} = X_{-j} \; 10^{B_j} \, T^{N_j} \exp\left[-\frac{T_j}{T} - \sum_{i=1}^{NS} (a_{ij}' - a_{ij}'') \frac{g_i^o}{RT}\right] [RT]^{(n_j'' - n_j')}$$

$$(\rho\sigma_m)^{\overline{a}_j} \prod_{k=1}^{NS} (\rho\sigma_k)^{a_{ij}''} \quad , \tag{4.10}$$

and the partial derivative coefficient of $\Delta \log T$ in Eq. (4.7) becomes

$$\frac{\partial f_i}{\partial \log T} = \sum_{j=1}^{JJ} (a_{ij}' - a_{ij}'') \left[(R_j - R_{-j})\left\{N_j + \frac{T_j}{T} - \overline{a}_j - n_j'\right\}\right.$$

$$\left. - R_{-j} \sum_{i=1}^{NS} (a_{ij}' - a_{ij}'') \frac{h_i}{RT}\right] \quad . \tag{4.11}$$

The functional for mixture reciprocal mole number, σ_m, is the same as for the equilibrium formulation, Eq. (3.16):

$$f_M \equiv \sum_{k=1}^{NS} \sigma_k - \sigma_m \tag{4.12}$$

and the corresponding correction equation is also the same as Eq. (3.33):

$$\sum_{k=1}^{NS} \sigma_k \, \Delta \log \sigma_k - \sigma_m \, \Delta \log \sigma_m = \sigma_m - \sum_{k=1}^{NS} \sigma_k \quad . \tag{4.13}$$

The functional for temperature (actually, for conservation of thermal energy) is the same as Eq. (3.14),

$$f_T \equiv -\frac{H^*}{RT} + \sum_{i=1}^{NS} \frac{\sigma_i h_i}{RT} + \frac{(Q/A_P)}{RT} , \tag{4.14}$$

and the corresponding correction equation

$$\sum_{k=1}^{NS} \frac{h_k \sigma_k}{RT} \Delta \log \sigma_k + \left[\sum_{i=1}^{NS} \frac{C_{p_i}}{R} \sigma_i + \frac{1}{RT} \frac{\partial (Q/A_P)}{\partial \log T} \right] \Delta \log T$$

$$= -\sum_{i=1}^{NS} \frac{\sigma_i h_i}{RT} - \frac{(Q/A_P)}{RT} + \sum_{i=1}^{NS} \frac{h_k^* \sigma_i^*}{RT} \tag{4.15}$$

4.2 *Solution of the Kinetic Correction Equations*

The basic approach is the same as described in Section 3 for the equilibrium correction equations. The details of the iteration procedure in Sec.3.3 and control of the underrelaxation parameter in Sec.3.4 are identical for kinetic stationary states as for equilibrium states. However, convergence control and the selection of estimates is somewhat more involved.

4.3 *Convergence*

The same convergence criterion ϵ is used as described in Sec.3.5. However, during the course of iterative solution of kinetic stationary states, oscillating nonconvergence is occasionally encountered [9], especially during ignition, when many species are undergoing relatively small changes, and the dominant chemical reactions are endothermic. One technique to avoid oscillation used by other investigators [9,10] is simply to decouple the energy equation; that is, to delete the energy equation and solve the remaining system of equations for assigned temperature, and then iteratively correct the assigned temperature until the desired value of enthalpy is achieved. This practice, while it effectively avoids the oscillation problem, is successful only at the expense of increased computer execution time. Solution with the fully coupled energy equation, as in the present formulation, is considerably faster. The occasional case of oscillating nonconvergence can be effectively dealt with by a partial or "one-way" decoupling of the energy equation, in which the variations of temperature are not permitted to affect the distribution of species mole numbers, but the mole number changes in each iteration are permitted to affect the temperature. This is easily achieved within the present formulation by setting the species-temperature partial derivatives, Eq. (4.7), equal to zero whenever oscillating convergence is encountered or suspected. The penalty is at worst a somewhat slower progress toward convergence than the fully coupled equations, although still considerably faster than full decoupling of the energy equation. By judicious use of partial decoupling, rapid, stable convergence can be achieved reliably over a wide range of fuel type, fuel-air ratio and cell time constant A_P.

4.4 *Initial Estimates*

The set of correction equations to be solved, Eqs (4.4), (4.13) and (4.15), are $(NS + 2)$ in number. This is unfortunately a considerably larger system than the $(NLM + 2)$ number of equations required to be solved for chemical equilibrium. Further, the mole numbers $(\sigma_i,\ i = 1, NS)$ are far more sensitive to variations in the cell time constant A_P and in T than are the smoothly-varying Lagrange multipliers π_i. Efficiency of convergence, as well as stability or freedom from divergence, therefore requires a choice of estimate set $(\sigma_i^{(o)}, i = 1, NS)$ and $T^{(o)}$ which is quite close to the final solution. The most obvious choice of estimates is simply the stored values from a previous superiteration of the flow field. However, in the initial stage of the field solution, these estimates may not be available, or may be inadequate due to the rapidly changing field values of A_P and of the thermochemical variables. Therefore, a strategy is needed to cope automatically with these "problem cells" whenever they arise. Fortunately, experience with related problems in a homogeneous perfectly-stirred reactor code [11] is applicable here.

An example of a "problem cell" is a cell at which ignition occurs; that is, a cell which has largely "cold" or unreacted values of (σ_i^*) at upstream neighbouring nodes, but which will ultimately have a "hot" or (partially) reacted set of values T and (σ_i) at a stable solution condition. Possible solution states are illustrated in Fig.2. If adiabatic mixing (no reaction) is assumed, the curve "$A_P \to \infty$" (zero cell residence time), which represents the variation of H with T for the inlet mixture set $(\sigma_i^*,\ i = 1,\ NS)$, is also the solution set. The intersection of this curve with the constant inlet enthalpy $H = H^*$ line determines the solution temperature, which is equal to the effective inlet temperature, T^*. If chemical equilibrium is assumed, then the curve "$A_P = 0$" (infinite cell resi-

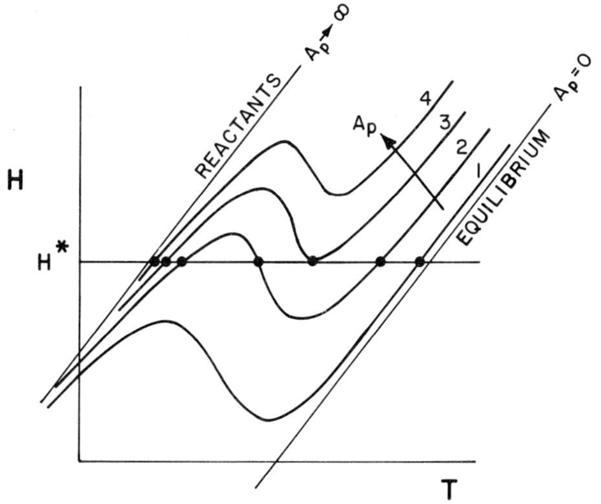

Fig.2 Schematic enthalpy-temperature diagram for chemical equilibrium and kinetic stationary states in a computational cell [10]. $Q = 0$ is assumed for clarity. Parametric curves are for different values of cell convection/diffusion coefficient A_P.

dence time) corresponds to the set of mole numbers which minimize the mixture Gibbs function at each temperature. The intersection of the "$A_P = 0$" curve with $H = H^*$ determines the solution set (σ_i^{eq}, $n = 1, NS$) and the equilibrium temperature T^{eq}.

For small values of A_P, such as represented by curve 1, which represent large but finite values of cell residence time, the solution point represented by the intersection of curve 1 with $H = H^*$ is unique, and corresponds to values of (σ_i, $i = 1, NS$) and T which differ only slightly from the equilibrium values; obviously the equilibrium solution is an excellent estimate set for this kinetic stationary state. Increasing values of A_P (decreasing cell residence time) correspond to non-equilibrium solutions which are further away from equilibrium.

To complicate matters further, it is possible for multiple solutions to exist, as illustrated by curve 2. Of the three intersections of this curve with $H = H^*$, the lower and upper temperature values are physically stable stationary states, which correspond to slow oxidation and to combustion, respectively [10]. The central branch is physically unstable, but satisfies all of the conservation equations. If an "upper branch" or combustion solution is desired, estimates must obviously be chosen so that the desired state is approached from higher temperature stationary states, that is, for values of A_P less than that of curve 2.

Curve 3 illustrates an upper branch solution which represents incipient extinction or "blowout"; an increase in A_P such as from curve 3 to 4 would cause a unique, lower-branch solution state to result. If the final solution for the cell should correspond to this condition, it is likely that intermediate solutions obtained for this cell during "super-iteration" of the field solution may vary between upper-branch (combustion) and lower-branch (slow oxidation) solutions. A successful computational strategy therefore requires that, if convergence is not obtained from previous solution values as estimates within a very few iterations, an upper-branch solution should be approached from the right (high temperature); that is, from an equilibrium estimate, with systematically increasing values of the parameter A_P up to the data value. This strategy is necessary to avoid having a cold, non-reacting solution propagate throughout the field when a combustion solution is desired. It is important to understand that the "cold" or lower-branch solution is as physically meaningful as is the combustion solution. In actual furnaces and combustors, for example, inflow of a given mixture of reactants does not automatically guarantee combustion: some method of ignition such as a continuous igniter or pilot flame, or self-ignition by convective or diffusive recirculation of burned products, is essential. These computational difficulties, and the technique for their solution, simply reflect the actual physical processes which occur in a self-stabilized flame, in a finite volume as well as in a computational cell embedded in a combustion flow field.

5. A DEMONSTRATION CALCULATION

While the computational scheme outlined in previous chapters was designed for use with flows of either parabolic or elliptic character, the principles of its application are most easily demonstrated in calculation of a parabolic reacting flow.

The scheme for calculation of chemical equilibrium and kinetic stationary states has been coded and documented as a collection of FORTRAN-IV subprograms collectively entitled CREK [12], for "Combustion Reaction Equilibrium and Kinetics". Modifications were made to an existing parabolic flow code, GENMIX-4A [1], in order to interface with CREK. These calculations were performed solely as a demonstration of the feasibility of the present computational scheme. Because of the various simplifying assumptions made for purposes of demonstration, validation by comparison with experimental measurements is not relevant, and was not attempted.

The problem considered was the thermally hypergolic ignition and combustion of two co-flowing streams of fuel and air undergoing turbulent shear mixing, as illustrated in Fig.3.

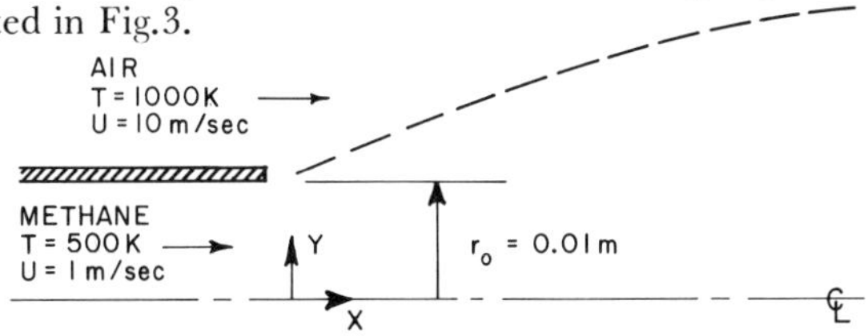

Fig.3 Axisymmetric shear flow mixing and combustion of initially unmixed fuel and air.

GENMIX-4A, being a code for solution of parabolic or "upstream-to-downstream" flows, utilizes a "six-node" version of Eq. (2.4), as follows:

$$(A_P - C)\,\phi_P = A_N\,\phi_N + A_S\,\phi_S + [A_{NW}\,\phi_{NW} + A_W\,\phi_W + A_{SW}\,\phi_{SW} + B]\ . \tag{5.1}$$

Since the flow direction is from left to right or from W to E in Fig.1, the values of the variable of interest ϕ, as well as the convection/diffusion coefficients A_d, are known at the upstream nodes (W, NW and SW), and GENMIX solves for simultaneous values of ϕ at the downstream nodes (P, N and S) by means of the tri-diagonal matrix algorithm. Field values of axial velocity u at each downstream station are thus solved by GENMIX in a non-iterative manner, by sweeping across the computational grid for simultaneous values of u at each station, and marching downstream after each cross-stream sweep. GENMIX utilizes a computationally efficient "expanding grid", so that the computational field occupies the entire shear layer as it grows wider downstream. Turbulent mixing is modelled in GENMIX by a conventional mixing-length/eddy viscosity model.

As modified for the present application, after the convection/diffusion coefficients A_d in Eq. (5.1) are established by GENMIX at each axial station for calculating the downstream values of u, the field is swept cross-stream, point by point, and CREK is called at each point to calculate simultaneously the local values of the thermochemical variables (σ_i, $i = 1$, NS) and T. CREK requires however that, at each node point of interest P, the thermochemical variables at adjacent downstream nodes N and S must be taken as fixed values, as discussed in Section 2. Cross-stream iteration is therefore required to converge the solution for the thermochemical fields at each axial station prior to marching downstream. The thermochemical field was taken as converged at each downstream step when the temperature correction at every cross-stream node was reduced to less than 1°C. All calculations were made on an IBM 360/67 computer employing the computationally efficient FORTRAN-H compiler.

5.2 Methane-air Mixing and Combustion

The thermochemical data for this system were taken from Gordon and McBride [8]. A reaction mechanism consisting of 18 species and 26 reactions (forward and reverse rates both considered) was utilized, as given in Table 1. A

Table 1. Methane combustion and NO_x formation mechanism.

Forward Rate Constant = $k_f = 10^{B_j} T^{N_j} \exp(-T_j/T)$. SI Units: kg-mol, m³, sec, deg K.

				$\rightleftharpoons$			B_j	N_j	T_j	Ref.
1.	CHO		M	CO	H	M	17.398	−1.5	8460.0	13
2.	CHO	H		CO	H_2		7.477	1.0	0.0	13
3.	CH_2O	O		CHO	OH		8.301	1.0	2215.0	13
4.	CHO	OH		CO	H_2O		7.477	1.0	0.0	13
5.	CHO	O		CO	OH		8.477	1.0	252.0	13
6.	CH_3	O		CH_2O	H		9.301	0.5	−151.0	13
7.	CH_4	O		CH_3	OH		7.0	1.0	4028.0	13
8.	CH_4	H		CH_3	H_2		7.700	1.0	5035.0	13
9.	CH_4	OH		CH_3	H_2O		10.477	0.0	2518.0	13
10.	CHO	O_2		CO	HO_2		9.903	0.0	0.0	13
11.	CO	OH		CO_2	H		6.602	0.5	0.0	13
12.	CO_2		M	CO	O	M	12.000	0.0	50353.0	13
13.	H	OH		H_2	O		6.903	1.0	3525.0	13
14.	H_2O		M	OH	H	M	12.477	0.0	52870.0	13
15.	H	HO_2		OH	OH		11.3981	0.0	957.0	13
16.	OH	H_2		H	H_2O		10.3980	0.0	2618.0	13
17.	H	O	M	OH		M	9.903	0.0	0.0	13
18.	OH	O		H	O_2		10.398	0.0	0.0	13
19.	H	O_2	M	HO_2		M	9.176	0.0	503.5	13
20.	OH	OH		H_2O	O		9.778	0.0	503.5	13
21.	OH	N		H	NO		8.778	0.5	4028.0	13
22.	H	N_2O		OH	N_2		10.903	0.0	7553.0	13
23.	N	NO		N_2	O		10.1760	0.0	0.0	13
24.	N	O_2		NO	O		6.778	1.0	3172.0	13
25.	N_2O	O		NO	NO		11.000	0.0	15000.0	13
26.	N_2O		M	N_2	O	M	11.000	0.0	25176.0	13

cross-stream grid of $N = 9$ was used for this demonstration calculation. Downstream integration was continued for 500 steps. Calculations were made for two cases, chemical equilibrium and finite-rate chemistry. In both cases, turbulent micromixing was taken to be infinite-rate; that is, the contact indices X_j and X_{-j} were taken as unity for all JJ reactions. The resulting temperature profiles are shown in Fig.4.

Inspection of the chemical equilibrium temperature profile in Fig.4 shows dramatically the effect of assuming both infinite-rate micromixing and infinite-rate chemical reaction. This assumption of "if it (macro)mixes, it burns" ignores

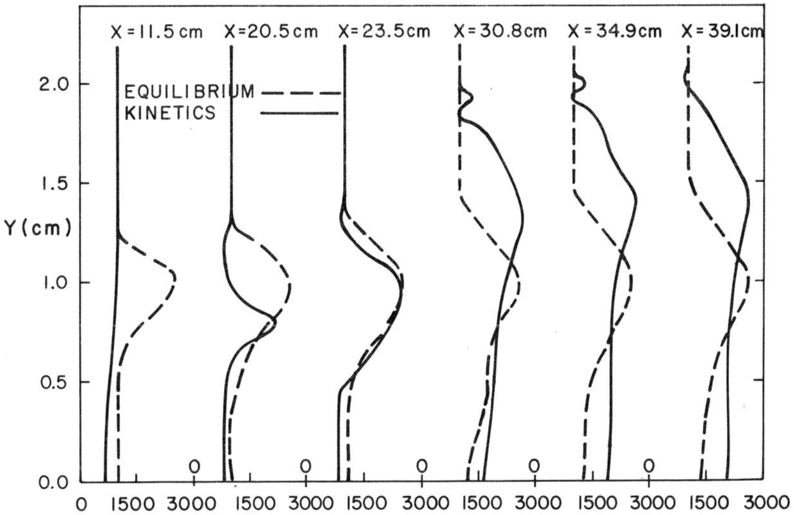

Fig. 4 Temperature (K) profiles at various axial locations, for CH_4-air mixing and combustion at conditions of Fig. 3.

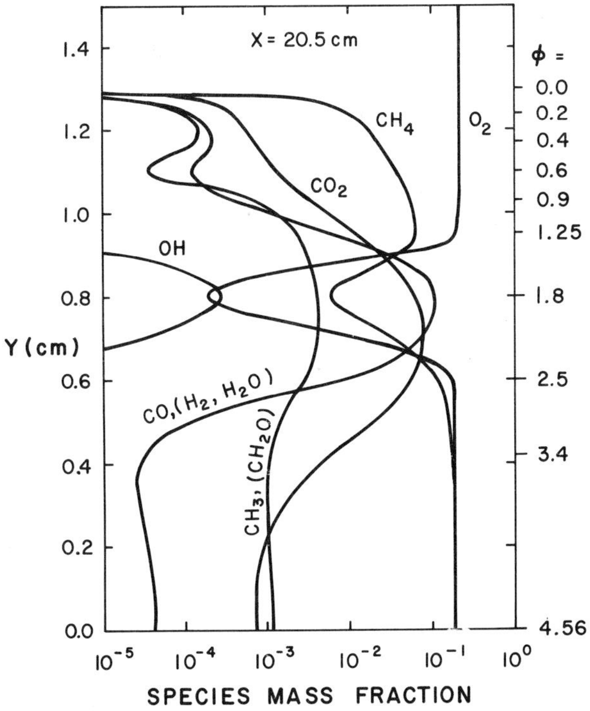

Fig. 5 Species mass-fraction profiles, at post-ignition station $x = 20.5$ cm for CH_4/air combustion at conditions of Fig. 3. Parentheses indicate similar behaviour for those species. Right ordinate is fuel/air equivalence ratio.

both the finite rate of diffusive mixing at the molecular level, and the need for the presence of either chemically reactive species or thermal sources to cause ignition to occur. Compared to the chemical equilibrium solution, the finite-rate or kinetic solution shows clearly the ignition delay resulting from the endothermic, oxidative pyrolysis of methane before the exothermic combustion of CO to CO_2, and of H_2 to H_2O, can occur. As the deflagration wave or flame front spreads outward from the ignition station at approximately $x = 20.5$ cm, the wrinkled temperature profile at the outer edge of the shear layer indicates the region in which pyrolysis occurs.

Figure 5 presents the cross-stream profiles of major species of interest at the first station downstream of the ignition station, $x = 20.5$ cm. It may be seen that, prior to ignition, oxygen has already diffused from the free stream into the jet core, and fuel has diffused outward. The resulting post-ignition "hole" in the O_2 and CH_4 concentration profiles is correlated with a simultaneous increase in the concentration of OH, the primary intermediate species in both oxidating pyrolysis and combustion. The folded profile predicted for some species near the edge of the shear layer is not an artifact of the discretized solution scheme, but is due to the effect of the endothermic chemical reactions in the pyrolysis region near the leading edge of the outward-spreading combustion wave. It is interesting to note that the predicted rate of spread of the shear layer is greater for the kinetic case, following ignition, than for the equilibrium case.

Cross-stream profiles of concentration of the air pollutant NO and of oxygen atom O are presented in Fig.6, for the same post-ignition station at $x = 20.5$ cm as in Fig.5. The predicted values of NO concentration show a strong correlation with O concentrations, as expected from the assumed reaction mechanism. By comparison with the corresponding temperature profile at $x = 20.5$ cm, in Fig.4, it can be seen that the correlation of NO concentration with temperature is not as strong. In the kinetic case, the slow destruction rates of NO result in accumulation of NO in the jet centre due to diffusion inward from the flame front, whereas chemical equilibrium requires that the NO be reduced in the fuel-rich jet core.

In this calculation, typically two to five cross-stream iterations were required to converge the thermochemical field. This number was greatest at the locations immediately downstream of ignition, where the largest cross-stream gradients occurred, and least at downstream locations where cross-stream gradients had become less severe.

5.3 Hydrogen-air Combustion

For these computations a mechanism of 12 species and 14 reactions was considered, as given in Table 2. A cross-stream grid spacing of $N = 19$ was utilized, and the reaction contact indices X_j and X_{-j}, $j = 1, JJ$, were again taken to be unity, as for the methane-air calculations. Calculations with the H_2-air system were made because the kinetics are known to be very fast at the temperatures considered, so that it was anticipated that the kinetic solution would differ only slightly, if at all, from the chemical equilibrium solution. This was in fact found to be the case; unlike hydrocarbon fuels, hydrogen/air combustion agrees very

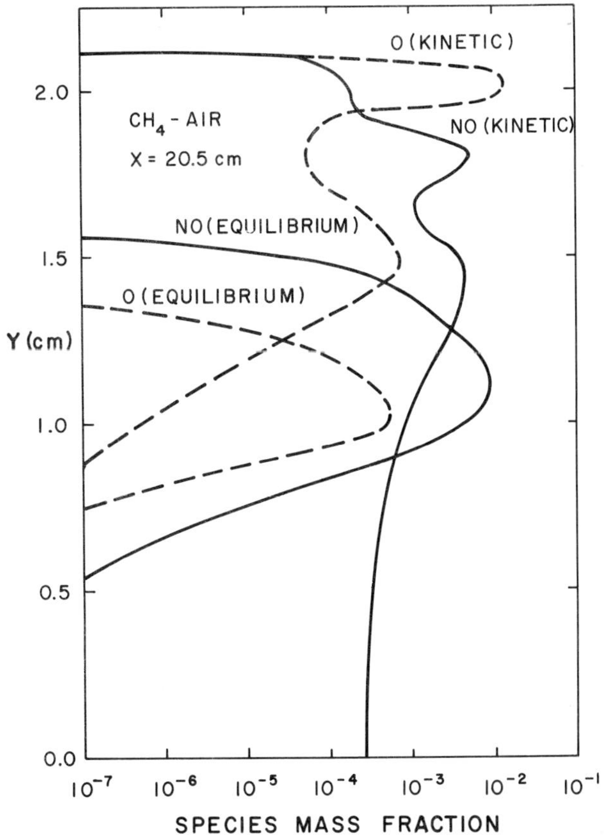

Fig. 6 Nitric oxide and oxygen-atom mass fraction profiles at post-ignition station $x = 20.5$ cm, for CH_4/air mixing and combustion at conditions of Fig. 3.

well with the "if it mixes, it burns" assumption if the initial temperature is sufficiently elevated. The resulting profiles are not presented here, as they are relatively uninteresting.

Comparative calculations were made of mixing with chemical reaction (kinetic or equilibrium, in this case) and without reaction, simply assuming adiabatic cross-stream mixing. The resulting mixture fraction profiles demonstrated the qualitative effect of decreasing mass density on transverse mixing: the predicted rate of spread of the chemically reacting, heat-releasing shear layer was about 1.5 times that of the non-reacting shear layer.

Also of interest was the fact that a greatly increased number of cross-grid iterations were required to converge the temperature profile, compared with the methane-air system; comparative data are provided in Table 3. More than ten such iterations were often required, owing to the very steep gradients resulting from the "if it mixes, it burns" assumption. The infinite-rate micromixing assumption thus resulted in a very great increase in execution time. In spite of the additional complexity in formulating and including a finite micromixing model (which

Table 2. Hydrogen combustion and NO_x formation mechanism.

Forward Rate Constant = $k_f = 10^{B_j} T^{N_j} \exp(-T_j/T)$. SI Units: kg-mol, m³, sec, deg K.

				$\rightleftharpoons$			B_j	N_j	T_j	Ref.
1.	H	O_2		O	OH		17.342	0.0	8450.0	14
2.	O	H_2		H	OH		13.255	1.0	4480.0	14
3.	OH	H_2		H_2O	H		16.342	0.0	2590.0	14
4.	H_2O	O		OH	OH		16.833	0.0	9240.0	14
5.	H	OH	M	H_2O		M	28.845	−2.0	0.0	14
6.	H_2	O_2		HO_2	H		16.740	0.0	29100.0	14
7.	H	HO_2		OH	OH		17.398	0.0	950.0	14
8.	H	O_2	M	HO_2		M	21.602	0.0	−500.0	14
9.	OH	N		H	NO		8.778	0.5	4028.0	13
10.	H	N_2O		OH	N_2		10.903	0.0	7553.0	13
11.	N	NO		N_2	O		10.176	0.0	0.0	13
12.	N	O_2		NO	O		6.778	1.0	3172.0	13
13.	N_2O	O		NO	NO		11.000	0.0	15000.0	13
14.	N_2O		M	N_2	O	M	11.000	0.0	25176.0	13

Table 3. Comparison of solution parameters for fuel-air mixing and combustion in a turbulent shear flow, as shown in Fig.3. Calculations performed on an IBM 360/67 computer with FORTRAN-H compiler.

	Fuel	
	CH_4	H_2
Mechanism		
Number of species, *NS*	18	12
Number of reactions, *JJ*	26	14
Cross-stream grid size, *N*	9	19
Average execution time per step: (sec)		
Mixing only	0.3 sec	0.18 sec
Equilibrium reaction	1.25	1.76
Kinetic reaction	5.0	14.1

would presumably lead to values of contact indices X_j between near-zero and, perhaps, ten) we may expect that the computational time would be reduced because of the less severe transverse temperature and species concentration gradients resulting from the damping action of finite-rate micromixing on the chemical heat release rate.

6. SUMMARY AND CONCLUSIONS

The most economical, finite-difference schemes for solving hydrodynamic fields (that is, the field differential equations governing the flow of fluids) proceed by solution for one variable at a time simultaneously at many grid points in the computational field. In contrast, the best way of solving the thermochemical field is optimal when simultaneous values of *all* variables are obtained at individual points in the computational field. This reasoning led to a proposal to solve chemically reacting flows by using two separate solution algorithms in "super-iteration", by (1) assuming the thermochemical variables as fixed, and solving for the hydrodynamic field one variable at a time, and (2) taking the hydrodynamic variables as fixed, and solving simultaneously for the thermochemical variables at each point in the field, and then repeating steps (1) and (2) until a converged solution for both fields is obtained. While only application to a chemically reacting parabolic or "upstream-to-downstream" flow has been demonstrated here, the technique has been applied with similar success to a highly turbulent, elliptic flow with strong convective recirculation [15].

The very important and difficult problem of calculating chemical reaction rates when both finite-rate chemistry and finite-rate turbulent micromixing are rate-influencing has only been touched on. The solution scheme proposed provides for a "contact index" correction term to the homogeneous, Arrhenius-type reaction rate expressions, but the physical correctness of this approach has not been demonstrated. However, experience with the demonstration calculations for mixing and combustion of two co-flowing streams of fuel and air suggest that the inclusion of corrections to the contact indices for finite-rate micromixing will probably cause only a very modest increase in calculation time, and may actually decrease the solution time, due to the damping effect of reduced mixing rates on predicted field gradients, which in turn leads to more rapid convergence of the "super-iteration" procedure.

Finally, a word or two of explanation should be given to rationalize the practice of ignoring the differences in convective/diffusive and chemical-kinetic time scales. First, the use of classical, one-dimensional chemical-kinetic integrators for steady flows (by time-integrating to the steady state) is completely avoided by the present technique, as the steady-state solution is independent of the path by which it is attained. For transient solutions, the present technique provides unconditional stability with respect to any chosen time step, without "time-splitting", at the sole expense of accuracy. However, as pointed out by Young and Boris [16], there is no point in seeking accuracy for the thermochemical variables in a transient reactive flow greater than that of the hydrodynamic solution; efficiency and convergence stability are paramount considerations. The present technique appears to satisfy both these criteria, while permitting any degree of accuracy the user is willing to pay for in terms of execution time.

ACKNOWLEDGEMENTS

Computing time for the demonstration calculations was made available by the Computing Center at Washington State University, and valuable programming assistance was provided by colleagues D.E. Stock and J.J. Wormeck. The basic solution scheme was developed while the author was a visitor at Imperial College of Science and Technology, University of London, during 1974-75. Financial assistance by Washington State University, and by the U.S.-U.K. Educational Commission in the form of a Fulbright-Hays research grant, is gratefully acknowledged.

REFERENCES

1. Spalding, D.B. "A General Computer Program for Two-Dimensional Boundary Layer Problems", Report No. HTS/73/48, Imperial College of Science and Technology, London, 1973.
2. Gosman, A.D. and Pun, W.M. "Documentation for TEACH, a Program for Calculation of Laminar or Turbulent Elliptic Flows", Department of Mechanical Engineering, Imperial College, London, 1973.
3. Patankar, S.V. "Numerical Prediction of Turbulent Flows". *In* "Studies in Convection", **1**, 1, 1975.
4. Borghi, R. "Chemical Reaction Calculations in Turbulent Flows". *In* "Advances in Geophysics", **Vol.18B**, Academic Press, 1974.
5. Donaldson, C.DuP. and Hilst, G.R. "Chemical Reactions in Inhomogeneous Mixtures: The Effect of Turbulent Mixing", *Proceedings of the 1972 Heat Transfer and Fluid Mechanics Institute*, R.B. Landis and G.J. Hordemann (eds), Stanford University Press, 1972.
6. Anasoulis, R.F. and McDonald, H. "A Study of Combustor Flow Computations and Comparison with Experiment", Report EPA-650/2-73-045, U.S. Environmental Protection Agency, 1973.
7. Kennedy, L.A. and Scaccia, C. "Modelling of Combustion Chambers for Predicting Pollutant Concentrations", ASME Paper 73-HT-22, 1973.
8. Gordon, S. and McBride, B. "Computer Program for Calculation of Complex Chemical Equilibrium Compositions", NASA SP-273, 1971.
9. Osgerby, I.T. "An Efficient Numerical Method for Stirred Reactor Calculations", Arnold Engineering Development Center Report AEDC-TR-72-164 (and AFOSR-TR-72-0910), 1972.
10. Jones, A. and Prothero, A. "The Solution of the Steady-State Equations for an Adiabatic Stirred Reactor", *Comb. & Flame* **12**, 5, 457, 1968.
11. Pratt, D.T. "PSR – A Computer Program for Calculation of Steady-Flow, Homogeneous Combustion Reaction Kinetics", Bulletin 336, Washington State University College of Engineering, 1974.
12. Pratt, D.T. and Wormeck, J.J. "CREK: A Computer Program for Calculation of Combustion Reaction Equilibrium and Kinetics in Laminar or Turbulent Flows", Washington State University Report WSU-ME-TEL-76-1, 1976.
13. Waldman, C.H., Wilson, R.P., Jr. and Maloney, K.L. "Kinetic Mechanisms of Methane-Air Combustion with Pollutant Formation", Environmental Protection Technology Series EPA-650/2-74-045, 1974.
14. Baulch, D.L., Drysdale, D.D., Horne, D.G. and Floyd, A.C. "Evaluated Kinetic Data for High-Temperature Reactions", **Vol.2**, Chemical Rubber Co. Press, Cleveland, Ohio, 1973.
15. Wormeck, J.J. and Pratt, D.T. "Computer Modelling of Combustion in a Longwell Jet-stirred Reactor", *Sixteenth Symposium (International) on Combustion*, The Combustion Institite, 1977.
16. Young, T.R. and Boris, J.P. "A Numerical Technique for Solving Stiff Ordinary Differential Equations Associated with Reactive Flow Problems", U.S. Naval Research Laboratory Memorandum Report 2611, 1973.

NOMENCLATURE

A_E, A_W, A_N A_S, A_H, A_L	Combined convection and diffusion coefficients
A_P	$(A_E + A_W + A_N + A_S + A_L + A_{PP} - C)$, etc.
A_{PP}	Previous time step influence coefficient
a_{ij}^L	Number of kg-atoms of element i per kg-mole of species j
B	Constant in linearized source term
B_j, B_{-j}	Exponent in modified Arrhenius pre-exponential factor
b_i	kg-atom number of element i in mixture

b_i^*	Average b_i of adjacent nodes, weighted by respective A_d's
C	Coefficient of linearized non-chemical-kinetic source term
C_{P_i}	Constant-pressure specific heat capacity of ideal gas species i
g_k^o	$h_k - Ts_k^o$; one-atmosphere ideal-gas partial molal specific Gibbs function of species k
g_k	$h_k - Ts_k^o + RT\ [\log(\sigma_k/\sigma_m) + \log(P/P_o)]$: ideal-gas partial molal specific Gibbs function of species k
H_d	Mixture specific enthalpy at neighbour nodes E, W, N, S, H, L and PP
H_P^*	Average of mixture specific enthalpy at neighbour nodes, with A_d's as weighting coefficients
h_k	Ideal-gas enthalpy of species k
JJ	The number of distinct chemical reactions in the kinetic mechanism under consideration
n_j', n_j''	$n_j' = \sum_{i=1}^{NS} a_{ij}'$; $n_j'' = \sum_{i=1}^{NS} a_{ij}''$: molecularity of forward and reverse reactions j, dimensionless
N_j, N_{-j}	Exponent of modified Arrhenius pre-exponential factor for forward and reverse reactions j
NLM	The number of distinct elements which comprise the NS chemical species under consideration
NS	The number of distinct chemical species under consideration
P	Pressure
P_o	Pressure of one standard atmosphere: 101,325.0 N m^{-2}
Q	Enthalpy heat loss term, $-S_H$
R	Universal gas constant: 8,314.3 J/(kg-mol)(K)
R_j, R_{-j}	Rate of forward and reverse reactions j, respectively
S_{σ_i}	Source term in finite-difference equation for conservation of species i
S_H	Source term in finite-difference equation for conservation of thermal energy
s_i^o	One-atmosphere ideal gas specific entropy of species i
s_i	Ideal gas specific entropy of species i
T	Temperature
T_j	Activation temperature (activation energy divided by gas constant) for reaction j in exponential term of modified Arrhenius rate "constant"
X_j, X_{-j}	Contact index of forward and reverse reactions j
a_{ij}', a_{ij}''	Stoichiometric coefficients for species i: number of (kg-moles)$_i$ in reaction j as a reactant or as a product, respectively
$\bar{a}_j$	Third-body stoichiometric coefficient in reaction j; either 0 or 1
δ_{ij}	Kronecker delta
ϵ	Convergence interval criterion
η	Underrelaxation factor
σ_i	Mole number of species i in mixture
$\sigma_{i,P}^*$	A_d-weighted average mole numbers at adjacent nodes

σ_m	$= \sum_{i=1}^{NS} \sigma_i$: reciprocal of mixture mean molecular weight
λ_i	Lagrange multipliers in Gibbs function minimization equations
π_i	Lagrange multipliers in Gibbs function minimization equations (non-dimensional)
ρ	Mixture ideal-gas mass density, $P/(RT\sigma_m)$

SUBJECT INDEX

J

K

L

M

N

P

R

S

T

U

V

W

Z